Pál Révész is a member of the Hungarian Academy of Sciences, and a professor of mathematics at the Technical University of Vienna and the Technical University of Budapest. He is also an adjunct professor of mathematics at Carleton University of Ottawa and József Attila University of Szeged (Hungary). He was an invited speaker at a number of international conferences on probability and statistics, as well as at the International Congress of Mathematicians in Helsinki, 1978.

In addition to publishing over 130 research papers, Prof Révész is the author of three books, *Law of Large Numbers* (Academic Press, 1968), *Strong Approximations in Probability and Statistics* (Academic Press, 1981; with M. Csörgő) and *Random Walk in Random and Non-random Environments* (World Scientific, 1990).

ADVANCED SERIES ON STATISTICAL SCIENCE & APPLIED PROBABILITY

Editor: Ole E. Barndorff-Nielsen

This Series comprises accounts, by leading experts, from those areas of statistical science and applied probability where advanced mathematical tools are essential. Both research monographs and textbooks will be included, and the series promotes purely theoretical works as well as studies that combine theory and applications.

*Published**

Vol. 22 *Stochastic Interest Rate Modeling with Fixed Income Derivative Pricing (Third Edition)*
by N. Privault

Vol. 21 *Change of Time and Change of Measure (Second Edition)*
by O. E. Barndorff-Nielsen and A. Shiryaev

Vol. 20 *Spatial Branching in Random Environments and with Interaction*
by J. Engländer

Vol. 19 *Risk-Sensitive Investment Management*
by M. H. A. Davis and S. Lleo

Vol. 18 *Analysis for Diffusion Processes on Riemannian Manifolds*
by F.-Y. Wang

Vol. 17 *Modeling and Pricing in Financial Markets for Weather Derivatives*
by F. E. Benth and J. Šaltytė Benth

Vol. 16 *An Elementary Introduction to Stochastic Interest Rate Modeling (Second Edition)*
by N. Privault

Vol. 15 *Hedging Derivatives*
by T. Rheinländer and J. Sexton

Vol. 14 *Ruin Probabilities (Second Edition)*
by S. Asmussen and H. Albrecher

Vol. 13 *Change of Time and Change of Measure*
by O. E. Barndorff-Nielsen and A. Shiryaev

*To view the complete list of the published volumes in the series, please visit:
http://www.worldscientific.com/series/asssap

RANDOM WALKS OF INFINITELY MANY PARTICLES

Pál Révész
Technical University of Vienna

World Scientific
Singapore • New Jersey • London • Hong Kong

Published by

World Scientific Publishing Co. Pte. Ltd.
5 Toh Tuck Link, Singapore 596224
USA office: 27 Warren Street, Suite 401-402, Hackensack, NJ 07601
UK office: 57 Shelton Street, Covent Garden, London WC2H 9HE

British Library Cataloguing-in-Publication Data
A catalogue record for this book is available from the British Library.

Advanced Series on Statistical Science and Applied Probability — Vol. 1
RANDOM WALKS OF INFINITELY MANY PARTICLES

ISBN-13 978-981-02-1784-6
ISBN-10 981-02-1784-6

Preface

My previous book "Random walk in random and non-random environments" (World Scientific, 1990) intended to give a comprehensive survey of the properties of a simple, symmetric random walk in $\mathbb{Z}^d$. Writing this book I realized that I could not present a survey that was complete or almost complete. In fact, in this very classical topic we experience the mushrooming of new puzzling problems every day. Hence I had two options: either I do not finish the writing of the book ever or I finish it at a point chosen arbitrarily and start to work on a new subject. As you see I have chosen the second option.

I realized that most of my friends had been dealing with random walks of many particles, moving non-independently i.e. with interactions among the particles. They used the book "Interacting particle systems" by Liggett as their bible. So, I decided that finishing my book, I should turn my attention to interacting particles as well. However, in the meantime I read a paper of J. Walsh [55] which dealt with the independent motions of infinitely many particles, considering also the case, when particles are multiplying according to the law of a branching process.

These ideas led me to study these kind of independent motions, before turning to the case of interacting particles. Hence the title of the present book could be "Non-interacting particle systems". However, branching is a kind of non-independence (interaction), therefore I gave the present title to the book, not claiming anything about the dependence or independence of the motions.

In addition to the influence of Walsh's paper I got many valuable advices from my friends. Here I mention only my visits at Carleton University (Ottawa, 1992) and University of Wisconsin (Wisconsin, Madison, 1993). At Carleton I was lucky enough to meet with and talk to D. Dawson and E. Perkins, while in Wisconsin I had pleasant discussions with P. Ney. They had a great influence on my work.

I have to extend my best thanks to my home institute, Technical University of Vienna, providing me with a Sabbatical to complete my book.

Vienna, 1994

P. Révész
Technical University of Vienna
Wiedner Hauptstraße 8–10/107
A–1040 Vienna, Austria

Contents

II. BRANCHING RANDOM WALK

III. STRASSEN TYPE THEOREMS

Introduction

The book consists of three parts.

In *Part* I the following model is treated. At time $t = 0$ in $\mathbb{R}^d$ there is a random point field. That is to say, there are countable, infinitely many points distributed according to a probability law. At $t = 0$ these points (particles) start to move independently, governed by the law of a stochastic process. Namely, the laws of the motions are equal to each other. Having this model we ask about the properties of the distribution of the particles at time T, when T is big. In order to expose the question more accurately, we have to tell:

1. What kind of law governs the initial random field.

2. What kind of law governs the motion of the particles.

3. What kind of limit properties are we interested in.

In Part I we consider mostly the case, when

1. The initial random field is a homogeneous Poisson field.

2. The motions are independent, standard Wiener processes.

3. We ask: in case $T \to \infty$,

   (i) how many points visit the unit-ball (around the origin) up to T,

   (ii) how many different points visit the unit-ball up to T,

   (iii) the maximum number of points located in the unit-ball at a time t $(0 \leq t \leq T)$.

It turns out that the role of the initial distribution is not very essential. Replacing the Poisson field by any "homogeneous enough" field the same results are valid. It is very likely that the Wiener process (which governs the motions) can be replaced by any "similar" process. However, this question is not completely settled.

Part II is devoted to branching random walks. In this model at time $t = 0$ we have some particles (one or more) in $\mathbb{R}^d$. These particles start to move independently and they die after a (random) time, but before their death they give birth to a few offsprings according to the law of a Galton - Watson process. The offsprings are moving independently according to the same law. We are interested in the distribution of the particles at time T, when T is big.

In order to explore the question more precisely we have to tell:

1. What kind of law governs the initial random field.
2. What kind of law governs the motions of the particles.
3. What is the law of the lifetime of a particle.
4. What kind of Galton - Watson process governs the multiplications.
5. What kind of limit properties are we interested in.

In Part II we consider mostly the case, when

1. The initial random field is a single particle located in $0 \in \mathbb{Z}^d$ at time $t = 0$.
2. The motions are simple, symmetric random walks in $\mathbb{Z}^d$
3. The lifetimes of the particles are equal to one.
4. The Galton - Watson process is supercritical.
5. The questions are:
   (i) the limit-behaviour of the number of particles located in a fixed $x \in \mathbb{Z}^d$,
   (ii) the limit-behaviour of the number of those particles which located at time T in a ball around the origin of radius R_T, where $R_T \nearrow \infty$ as $T \nearrow \infty$,
   (iii) the description of the domain, where the living particles are located at time T.

Part III is devoted to Strassen type theorems (functional laws of iterated logarithm). Instead of the original question of Strassen, when the path-properties of a single particle (governed by a Wiener process) are treated, here the path-properties of many independent Wiener processes are considered.

In the last twenty years a huge number of papers appeared on the above mentioned problems. However, due to the large numbers of possible models, it is very difficult to compare the results obtained. In the course of the present text we deal with those models which seem to be the simplest ones, without even mentioning the other models treated in the literature. Hence we give the references of those theorems only, which are given without proofs as tools for further results. Even in these cases, whenever it is possible, I refer to my book [44] instead of refering to the original papers.

We supply the reader with an appendix (Historical overview), where a collection of the most important relevant results is presented.

Notations and abbreviations

Notations

$\mathbb{R}^d$ is the d-dimensional Eucledean space

1. $x = (x_1, x_2, \ldots, x_d)$ is an element of $\mathbb{R}^d$ if $-\infty < x_i < \infty$ for any $i = 1, 2, \ldots, d$,
2. $\|x\|^2 = x_1^2 + x_2^2 + \cdots + x_d^2$,
3. $C(a, r) = \{x : \|x - a\| \le r\} \quad (r > 0)$,
4. $C(r) = C(0, r)$,
5. $|A|$ is the Lebesgue measure of A for any Borel set $A \subset \mathbb{R}^d$,
6. let

$$r_0 = r_0(d) = \begin{cases} 1/2 & \text{if} \quad d = 1, \\ \pi^{-1/2} & \text{if} \quad d = 2, \\ \pi^{-1/2}\left(\Gamma\left(\dfrac{d}{2} + 1\right)\right)^{1/d} & \text{if} \quad d \ge 3, \end{cases}$$

   then $|C(r_0)| = 1$ and let $C = C(r_0)$,

7. $\overline{A} = \mathbb{R}^d - A$,
8. $B(r, s) = C(r + s) - C(r) \quad (r, s > 0)$,
9. $e_1 = (1, 0, \ldots, 0), e_2 = (0, 1, \ldots, 0), \ldots, e_d = (0, 0, \ldots, 1)$ are the orthogonal unit-vectors of $\mathbb{R}^d$,
10. for any Borel set $A \subset \mathbb{R}^d$ let ∂A be the boundary of A,
11. $x = (x_1, x_2, \ldots, x_d) \le y = (y_1, y_2, \ldots, y_d)$ if $x_i \le y_i$ $(i = 1, 2, \ldots, d)$.

$\mathbb{Z}^d$ is the d-dimensional lattice

1. $x = (x_1, x_2, \ldots, x_d)$ is an element of $\mathbb{Z}^d$ if $x_i = 0, \pm 1, \pm 2, \ldots$ for any $i = 1, 2, \ldots, d$,
2. $|x| = |x_1| + |x_2| + \cdots + |x_d|$,
3. $C(a, r) = \{x : x \in \mathbb{Z}^d, \ |x - a| \le r\} \quad (r > 0)$,

4. $C(x) = C(x,1)$ i.e. $C(x)$ is the set of the neighbours of x.

5. $\#\{\dots\}$ is the cardinality of the set in brackets.

Note that the letter C is used in different ways in case $\mathbb{R}^d$ and in case $\mathbb{Z}^d$. In Part I resp. Part II the basic space is mostly $\mathbb{R}^d$ resp. $\mathbb{Z}^d$ and C is used accordingly.

Functions

1. $b_t = b(t) = (2t \log\log t)^{-1/2} \qquad (t > e^e)$,

2. $\gamma(t) = \gamma(t,a) = (2a(\log(t/a) + \log\log t))^{-1/2}$,

3. $\lfloor x \rfloor$ is the largest integer less than or equal to x,

4. $f(t) \gg g(t) \leftrightarrow g(t) = o(f(t)) \leftrightarrow \lim_{t\to\infty} \frac{f(t)}{g(t)} = \infty$,

5. $f(t) \sim g(t) \leftrightarrow \lim_{t\to\infty} \frac{f(t)}{g(t)} = 1$,

6. $g(t) = O(f(t)) \leftrightarrow 0 < \liminf_{t\to\infty} \frac{f(t)}{g(t)} \le \limsup_{t\to\infty} \frac{f(t)}{g(t)} < \infty$,

7. $\Phi(x) = \prod_{i=1}^{d} \frac{1}{\sqrt{2\pi}} \int_{-\infty}^{x_i} e^{-u^2/2} du \quad x = (x_1, x_2, \dots, x_d) \in \mathbb{R}^d$ is the standard normal distribution function,

8. $\Gamma(t) = \int_0^\infty x^{t-1} e^{-x} dx \quad (t > 0)$ is the gamma function,

9. $\log_p$ is the p-th iterated of log.

Probability

1. the random variable $N \in \mathcal{N}(m,\sigma)$ if and only if $\mathbf{P}\{\sigma^{-1}(N - m) < x\} = \Phi(x)$ $(\sigma > 0,\ -\infty < m < \infty)$,

2. π is a Poisson field on $\mathbb{R}^d$ (cf. Section 1.1),

3. $\{W(t),\ t \ge 0\}$ is a Wiener process in $\mathbb{R}^d$,

4. $S_n = S(n)$ is a simple, symmetric random walk in $\mathbb{Z}^d$, i.e. $S_0 = 0$ and $S_{n+1} - S_n = X_n$ is a sequence of independent, identically distributed random vectors with

$$\mathbf{P}\{X_n = e_i\} = \mathbf{P}\{X_n = -e_i\} = \frac{1}{2d} \quad (i = 1, 2, \dots, d),$$

5. for any $u \in \mathbb{Z}^d$, $v \in \mathbb{Z}^d$, $t = 0, 1, 2, \ldots$ let

$$p(u \rightsquigarrow v, t) = \mathbf{P}\{S_{s+t} = v \mid S_s = u\},$$

6. D^*-sequence see p. 35,

7. D^*-process see p. 43.

Abbreviations

1. r.v. = random variable,

2. i.i.d.r.v.'s = independent, identically distributed r.v.'s,

3. LIL = law of iterated logarithm,

4. CLT = central limit theorem,

5. i.o. = infinitely often,

6. a.s. = almost surely.

"Oh, look how so many are nimbly dispersing
Over the gardens, across the fields,
And the boats on the river happily coursing;
How the wide stream flows, how the water yields!
And that last one setting off, almost foundering
Under its load;"

Goethe: Faust
Translated by D. Luke

Part I

RANDOM WALK OF A RANDOM FIELD

Chapter 1

Brownian motion of a Poisson field

1.1 Poisson field

A point process π in $\mathbb{R}^d$ is called a Poisson field of parameter λ ($\lambda > 0$) if

$$\mathbf{P}\{\pi(\mathcal{A}) = k\} = \frac{(\lambda|\mathcal{A}|)^k}{k!}\exp(-\lambda|\mathcal{A}|) \qquad (k = 0, 1, 2, \ldots)$$

and $\pi(\mathcal{A}_1), \pi(\mathcal{A}_2), \ldots, \pi(\mathcal{A}_n)$ ($n = 2, 3, \ldots$) are independent r.v.'s where $\pi(\mathcal{A})$ is the number of the points of the process π in $\mathcal{A}$, $\mathcal{A} \subset \mathbb{R}^d$ is Lebesgue measurable, $|\mathcal{A}|$ is the Lebesgue measure of $\mathcal{A}$ and $\mathcal{A}_1, \mathcal{A}_2, \ldots, \mathcal{A}_n$ are disjoint Lebesgue measurable subsets of $\mathbb{R}^d$. As it is well known

$$\mathbf{E}\pi(\mathcal{A}) = \text{Var}\,\pi(\mathcal{A}) = \lambda|\mathcal{A}|.$$

Here we present three simple lemmas.

LEMMA 1.1 *Let Z be a Poisson r.v. i.e.*

$$\mathbf{P}\{Z = k\} = \frac{\lambda^k}{k!}e^{-\lambda} \qquad (k = 0, 1, 2, \ldots;\ \lambda > 0)$$

and let $Y_1, Y_2, \ldots$ be i.i.d.r.v.'s, being independent from Z, with

$$\mathbf{P}\{Y_1 = j\} = q_j \qquad (j = 1, 2, \ldots)$$

where $q_j \geq 0$, $\sum_{j=1}^{\infty} q_j = 1$. Finally let

$$I_j(Y_i) = \begin{cases} 1 & \text{if} \quad Y_i = j, \\ 0 & \text{if} \quad Y_i \neq j \end{cases}$$

and

$$X_j = \sum_{i=1}^{Z} I_j(Y_i).$$

Then $X_1, X_2, \ldots$ are independent r.v.'s with

$$X_1 + X_2 + \cdots = Z$$

and

$$\mathbf{P}\{X_j = k\} = \frac{(\lambda q_j)^k}{k!}e^{-\lambda q_j}$$

($j = 1, 2, \ldots;\ k = 0, 1, 2, \ldots$).

Remark 1.1. The above lemma can be interpreted as follows. Consider Z particles. The i-th ($i = 1, 2, \ldots, Z$) particle will be painted by the j-th colour with probability q_j. Let X_j be the number of particles painted by the j-th colour. Then the X_j's are independent Poisson r.v.'s of parameters λq_j respectively.

Proof of Lemma 1.1. For any $\ell = 1, 2, \ldots$ we have

$$\begin{aligned}
&\mathbf{P}\{X_1 = k_1, X_2 = k_2, \ldots, X_\ell = k_\ell\} = \\
&= e^{-\lambda} \sum_{j=K}^{\infty} \frac{\lambda^j}{j!} \mathbf{P}\{X_1 = k_1, X_2 = k_2, \ldots, X_\ell = k_\ell \mid Z = j\} = \\
&= e^{-\lambda} \sum_{j=K}^{\infty} \frac{\lambda^j}{j!} \frac{j!}{k_1! k_2! \ldots k_\ell! (j-K)!} q_1^{k_1} q_2^{k_2} \ldots q_\ell^{k_\ell} (1-Q)^{j-K} = \\
&= \frac{e^{-\lambda} \lambda^K}{k_1! k_2! \ldots k_\ell!} q_1^{k_1} q_2^{k_2} \ldots q_\ell^{k_\ell} \sum_{j=K}^{\infty} \frac{\lambda^{j-K}}{(j-K)!} (1-Q)^{j-K} = \\
&= \frac{e^{-\lambda Q} \lambda^K}{k_1! k_2! \ldots k_\ell!} q_1^{k_1} q_2^{k_2} \ldots q_\ell^{k_\ell} = \prod_{i=1}^{\ell} \frac{(\lambda q_i)^{k_i}}{k_i!} e^{-\lambda q_i}
\end{aligned}$$

where

$$K = K_\ell = k_1 + k_2 + \cdots + k_\ell, \qquad Q = Q_\ell = q_1 + q_2 + \cdots + q_\ell.$$

Hence we have the Lemma.

Since the probability that any two points of π lie exactly the same distance from the origin is 0, we can order them by magnitude and shall denote them by $P_1, P_2, \ldots$ i.e. $\pi = \{P_1, P_2, \ldots\}$.

The following lemma is a trivial consequence of Lemma 1.1.

LEMMA 1.2 *By the notations of Lemma* 1.1 *let*

$$\pi_j = \{P_\ell : Y_\ell = j\} \qquad (j = 1, 2, \ldots)$$

i.e. π_j *consists of those* P_ℓ *elements of* π *which are painted by colour* j. *Then* $\pi_1, \pi_2, \ldots$ *are independent Poisson fields of parameters* $\lambda q_1, \lambda q_2, \ldots$ *respectively (cf. Remark* 1.1.*).*

LEMMA 1.3 *Let* $x \in \mathbb{R}^d$ *be an arbitrary vector and let*

$$\pi + x = \{P_1 + x, P_2 + x, \ldots\}.$$

Then $\pi + x$ *is a Poisson field of parameter* λ *if* $\pi = \{P_1, P_2, \ldots\}$ *is a Poisson field of parameter* λ.

Proof is trivial.

THEOREM 1.1 *Let $X_1, X_2, \ldots$, $(X_i \in \mathbb{R}^d)$ be a sequence of i.i.d. random vectors and let $\pi = \{P_1, P_2, \ldots\}$ be a Poisson random field of parameter λ being independent from $\{X_1, X_2, \ldots\}$. Then $\{P_1 + X_1, P_2 + X_2, \ldots\}$ is a Poisson random field of parameter λ as well.*

Proof. If X_1 is discrete then our statement is a trivial consequence of Lemmas 1.2 and 1.3. Approximating X_1 by a discrete r.v. we obtain Theorem 1.1.

1.2 The model

Let $\pi = \{P_1, P_2, \ldots\}$ be a Poisson field of parameter λ in $\mathbb{R}^d$. Let $W_1(t), W_2(t), \ldots$ $(t \geq 0)$ be a sequence of independent $\mathbb{R}^d$ valued Wiener processes being also independent from π and define

$$P_i(t) = P_i + W_i(t) \qquad (i = 1, 2, \ldots)$$

and

$$P(t) = \{P_1(t), P_2(t), \ldots\}.$$

Clearly $P(0) = \pi$. By Theorem 1.1 we have

THEOREM 1.2 *For any fixed $t \geq 0$ the point process $P(t)$ is a Poisson field of parameter λ.*

Introduce the following notations.

(i) For any Borel set $A \subset \mathbb{R}^d$ and $t \geq 0$ let

$$P(A, t) = \{P_i(t) : P_i(t) \in A\}$$

i.e. $P(A, t)$ is a random subset of A (finite with probability 1 if $|A| < \infty$). In fact, it is the set of those points $P_i(t)$ which are located in A.

(ii) For any Borel set $A \subset \mathbb{R}^d$ and $t \geq 0$ let

$$s(A, t) = \#\{i : P_i(t) \in A\} = \#\{P(A, t)\}.$$

(iii) Let C be the ball around the origin with $|C| = 1$ and let

$$s(t) = s(C, t).$$

(iv)

$$S(T) = \sup_{0 \leq t \leq T} s(t).$$

(v) Let $N(\mathcal{A},t)$ resp. $N(t)$be the number of distinct points which visit $\mathcal{A}$ resp. $\mathcal{C}$ up to time t, i.e.

$$N(\mathcal{A},t) = \#\{i: \exists\, 0 \le s \le t \text{ such that } P_i(s) \in \mathcal{A}\}$$

resp.

$$N(t) = N(\mathcal{C},t).$$

Note that $N(0) = s(0)$.

(vi)

$$D(\mathcal{A},T) = \int_0^T s(\mathcal{A},t)dt$$

resp.

$$D(T) = D(\mathcal{C},T).$$

Note that $P(t)$ is a stationary, ergodic Markov process. $s(t)$ is also a stationary, ergodic process having Poisson distribution of parameter λ for any $t \ge 0$, however $s(t)$ is not a Markov process, not a martingale as well.

1.3 New particles in a ball

This Section is devoted to the study of the properties of $N(t)$. At first we prove that $N(t)$ is a Poisson r.v.

THEOREM 1.3

$$\mathbf{P}\{N(t) = k\} = \frac{\mu^k}{k!}e^{-\mu} \qquad (k = 0,1,2,\ldots) \tag{1.1}$$

where

$$\mu = \mu(t) = \mathbf{E}N(t) = \begin{cases} \lambda + 2\lambda\left(\frac{2}{\pi}\right)^{1/2}\int_1^\infty\int_{Rt^{-1/2}}^\infty e^{-u^2/2}dudR & \text{if } d = 1, \\ \lambda\left(\frac{2\pi t}{\log t} + (C_d + o(1))\frac{t}{(\log t)^2}\right) & \text{if } d = 2, \\ \lambda\left((6\pi^2)^{1/3}t + \left(\frac{3^4\cdot 2^7}{\pi}\right)^{1/6}t^{1/2}\right) & \text{if } d = 3, \\ \lambda(C_d + o(1))t & \text{if } d \ge 4 \end{cases} \tag{1.2}$$

where C_d $(d = 2,4,5\ldots)$ are positive constants.

Note that if $d=1$ then

$$\mu \sim 2\lambda\left(\frac{2}{\pi}\right)^{1/2}\int_1^\infty\int_{Rt^{-1/2}}^\infty e^{-u^2/2}dudR =$$
$$= 2\lambda\left(\frac{2t}{\pi}\right)^{1/2}\int_{t^{-1/2}}^\infty\int_v^\infty e^{-u^2/2}dudv \sim 2\lambda\left(\frac{2t}{\pi}\right)^{1/2}\int_0^\infty\int_v^\infty e^{-u^2/2}dudv =$$
$$= 2\lambda\left(\frac{2t}{\pi}\right)^{1/2}\int_0^\infty te^{-t^2/2}dt = 2\lambda\left(\frac{2t}{\pi}\right)^{1/2}$$

as $t \to \infty$.

Proof. Let $x \in \mathbb{R}^d$ with $\|x\| = R$ and

$$p(x,t) = p(R,t) = \mathbf{P}\{\exists\, 0 \le s \le t \text{ for which } P_i(s) \in C \mid P_i(0) = x\}$$

be the probability that a point starting from x visits C up to time t. Let

$$\mathcal{B} = \mathcal{B}(R, \Delta R) = \{x : \; R \le \|x\| \le R + \Delta R\}$$

and

$$\pi(R, \Delta R) = s(\mathcal{B}, 0).$$

Clearly

$$\mathbf{P}\{\pi(R,\Delta R) = k\} = \frac{(\lambda|\mathcal{B}|)^k}{k!}e^{-\lambda|\mathcal{B}|} \qquad (k = 0, 1, 2, \ldots)$$

where

$$|\mathcal{B}| = ((R+\Delta R)^d - R^d)\omega_d \sim d\omega_d R^{d-1}\Delta R$$

and ω_d is the volume of the ball in $\mathbb{R}^d$ of radius 1 i.e.

$$\omega_d = \begin{cases} 2 & \text{if } d = 1, \\ \pi & \text{if } d = 2, \\ \dfrac{\pi^{d/2}}{\Gamma(d/2+1)} & \text{if } d \ge 3. \end{cases}$$

For any $R \ge r_0$ let

$$N(R,t) = \#\{i : \; \exists\, 0 \le s \le t \text{ such that } P_i(s) \in C, \; P_i(0) \in \mathcal{B}\}$$

where

$$r_0^d\omega_d = |C| = 1 \qquad \text{i.e.} \qquad r_0 = \omega_d^{-1/d}.$$

Then clearly as $\Delta R \to 0$ we have

$$\mathbf{P}\{N(R,t) = k\} \sim \sum_{j=k}^\infty \frac{(\lambda|\mathcal{B}|)^j}{j!}e^{-\lambda|\mathcal{B}|}\binom{j}{k}(p(R,t))^k(1-p(R,t))^{j-k} =$$

$$= e^{-\lambda|B|}\frac{(\lambda|B|p(R,t))^k}{k!}\sum_{j=k}^{\infty}\frac{(\lambda|B|)^{j-k}}{(j-k)!}(1-p(R,t))^{j-k} =$$

$$= e^{-\lambda|B|}\frac{(\lambda|B|p(R,t))^k}{k!}\exp(\lambda|B|(1-p(R,t))) =$$

$$= \frac{\Lambda^k}{k!}e^{-\Lambda}$$

where

$$\Lambda = \lambda|B|p(R,t) \sim \lambda d\omega_d p(R,t)R^{d-1}\Delta R.$$

Hence

$$\mathbf{P}\{N(t)=k\} = \frac{\mu^k}{k!}e^{-\mu}$$

where

$$\mu = \mu(t) = \mathbf{E}N(t) = \lambda + \lambda d\omega_d\int_{r_0}^{\infty} R^{d-1}p(R,t)dR \tag{1.3}$$

and we have (1.1) of Theorem 1.3.

The proof of (1.2) in the case $d = 1$ is very simple. In fact we have

$$p(R,t) = \mathbf{P}\{\sup_{0\le s\le t} W(s) \ge R\} = 2\left(1-\Phi\left(\frac{R}{\sqrt{t}}\right)\right) = \sqrt{\frac{2}{\pi}}\int_{Rt^{-1/2}}^{\infty} e^{-u^2/2}du.$$

Hence we have (1.2) by (1.3) for $d = 1$. In case $d \ge 2$ (1.2) is proved in [51].

As a simple consequence of Theorem 1.3 we get

THEOREM 1.4

$$\lim_{t\to\infty}\mathbf{P}\left\{\frac{N(t)-\mathbf{E}N(t)}{(\mathbf{E}N(t))^{1/2}} < x\right\} = \Phi(x) \qquad (-\infty < x < \infty).$$

Theorem 1.3 suggests that $\{N(t),\ t \ge 0\}$ should be a Poisson process. In fact we have

THEOREM 1.5 *$\{N(t),\ t\ge 0\}$ is a Poisson process with $N(0) = s(0)$, i.e. for any $0 \le t_1 < t_2 \le t_3 < \ldots \le t_{2k-1} < t_{2k}$ the increments $N(t_{2i}) - N(t_{2i-1})$ $(i = 1,2,\ldots,k)$ are independent Poisson r.v.'s.*

The proof of this theorem is based on the following:

LEMMA 1.4 *Let $\overline{C} = \mathbb{R}^d - C$ be the complement of C and let $P^*(\overline{C},t)$ be the set of those points of $P(\overline{C},t)$ which did not meet C before t. Then for any $t > 0$ $\{N(s),\ 0 \le s \le t\}$ and $P^*(\overline{C},t)$ are independent.*

Proof. Let $R \le P_i(0) \le R + \Delta R$. Then the process $P_i(t)$ meets C up to time t with probability $p(R,t)$. Hence the points located in B at time $t = 0$ and not visiting C

up to time t form a Poisson field, independent from $\{N(s),\ 0 \le s \le t\}$ (cf. Lemma 1.1).

Consequently the Poisson field $P^*(\overline{C},t)$ is independent from $\{N(s),\ 0 \le s \le t\}$ and we have the Lemma.

Proof of Theorem 1.5. Since $N(t_{2k}) - N(t_{2k-1})$ depends on $P^*(\overline{C},t_{2k-2})$ and by Lemma 1.4 it is independent from $\{N(s),\ 0 \le s \le t_{2k-2}\}$ we have that $N(t_{2k}) - N(t_{2k-1})$ and $N(t_{2k-2})$ are independent Poisson r.v.'s. The general statement can be obtained similarly.

We prove two further theorems on the properties of $N(t)$. The first one is a strong invariance principle saying that $N(t) - \mathbf{E}N(t)$ can be approximated by a time changed Wiener process, the second one is a LIL.

THEOREM 1.6 *Without loss of generality we can assume that $N(t)$ is defined on a rich enough probability space. Then on the same space there exists a Wiener process $\{W(t),\ t \ge 0\}$ such that*

$$\sup_{0 \le t \le T} |(N(t) - \mathbf{E}N(t)) - W(\mu(t))| = O(\log T) \quad a.s.$$

Proof. Note that $\mu(t) = \mathbf{E}N(t)$ is a strictly increasing continuous function on $[0,\infty)$ with $\mu(0) = \lambda$. Hence

$$\tilde{N}(t - \lambda) = N(\mu^{-1}(t)) \qquad (t \ge \lambda)$$

is a homogeneous Poisson process with $\tilde{N}(0) = N(0) = s(0)$ ($\mu^{-1}(\cdot)$ is the inverse function of $\mu(\cdot)$ with $\mu^{-1}(\lambda) = 0$). Consequently (cf. [19] Theorem 2.6.1) there exists a Wiener process $\{W(t),\ t \ge 0\}$ such that

$$\sup_{0 \le t \le T} |\tilde{N}(t) - t - \lambda - W(t)| = O(\log \mu(T)) = O(\log T) \quad \text{a.s.}$$

Hence we have Theorem 1.6.

THEOREM 1.7

$$\limsup_{t\to\infty} \frac{N(t) - \mu(t)}{(2\mu(t)\log\log t)^{1/2}} = -\liminf_{t\to\infty} \frac{N(t) - \mu(t)}{(2\mu(t)\log\log t)^{1/2}} = 1 \quad a.s.$$

Proof. It is a trivial consequence of Theorem 1.6 and the LIL for Wiener process (cf. [44] Theorem 6.2).

For later reference we prove the following:

LEMMA 1.5 *Let*

$$t_i = \inf\{t : N(t) - N(0) \ge i\} \qquad (i = 1, 2, \ldots)$$

and α_i *be the solution of the equation*

$$\mathbf{E}N(\alpha_i) = \mu(\alpha_i) = i.$$

Then there exists a $C > 0$ *such that*

$$\mathbf{P}\left\{|t_i - \alpha_i| \geq y\frac{i^{1/2}}{\mu'(\alpha_i)}\right\} \leq \exp(-Cy^2)$$

if $y > 0$ *and* i *is big enough. Note that* $\mu(\cdot)$ *is a strictly increasing differentiable function having a nonincreasing bounded derivative.*

Proof. By Theorem 1.4 for $x > 0$ we have

$$\begin{aligned}
&\mathbf{P}\{|t_i - \alpha_i| \geq x\alpha_i^{1/2}\} = \\
&= \mathbf{P}\{t_i \geq \alpha_i + x\alpha_i^{1/2}\} + \mathbf{P}\{t_i \leq \alpha_i - x\alpha_i^{1/2}\} = \\
&= \mathbf{P}\{N(\alpha_i + x\alpha_i^{1/2}) - N(0) \leq i\} + \mathbf{P}\{N(\alpha_i - x\alpha_i^{1/2}) - N(0) \geq i\} = \\
&= \mathbf{P}\left\{\frac{N(\alpha_i + x\alpha_i^{1/2}) - N(0) - \mu(\alpha_i + x\alpha_i^{1/2})}{(\mu(\alpha_i + x\alpha_i^{1/2}))^{1/2}} \leq \frac{i - \mu(\alpha_i + x\alpha_i^{1/2})}{(\mu(\alpha_i + x\alpha_i^{1/2}))^{1/2}}\right\} + \\
&\quad + \mathbf{P}\left\{\frac{N(\alpha_i - x\alpha_i^{1/2}) - N(0) - \mu(\alpha_i - x\alpha_i^{1/2})}{(\mu(\alpha_i - x\alpha_i^{1/2}))^{1/2}} \geq \frac{i - \mu(\alpha_i - x\alpha_i^{1/2})}{(\mu(\alpha_i - x\alpha_i^{1/2}))^{1/2}}\right\} \sim \\
&\sim \Phi\left(\frac{i - \mu(\alpha_i + x\alpha_i^{1/2})}{(\mu(\alpha_i + x\alpha_i^{1/2}))^{1/2}}\right) - \Phi\left(\frac{i - \mu(\alpha_i - x\alpha_i^{1/2})}{(\mu(\alpha_i - x\alpha_i^{1/2}))^{1/2}}\right) + 1.
\end{aligned}$$

Since

$$\mu(\alpha_i - x\alpha_i^{1/2}) = \mu(\alpha_i) - x\alpha_i^{1/2}\mu'(\xi_i) = i - x\alpha_i^{1/2}\mu'(\xi_i)$$

with some $\alpha_i - x\alpha_i^{1/2} \leq \xi_i \leq \alpha_i$, choosing

$$x = y\left(\frac{i}{\alpha_i}\right)^{1/2}\frac{1}{\mu'(\alpha_i)}$$

we have

$$\begin{aligned}
\frac{i - \mu(\alpha_i - x\alpha_i^{1/2})}{(\mu(\alpha_i - x\alpha_i^{1/2}))^{1/2}} &= \frac{x\alpha_i^{1/2}\mu'(\xi_i)}{(i - x\alpha_i^{1/2}\mu'(\xi_i))^{1/2}} \geq \\
&\geq \frac{x\alpha_i^{1/2}\mu'(\xi_i)}{i^{1/2}} \geq x\mu'(\alpha_i)\left(\frac{\alpha_i}{i}\right)^{1/2} = y.
\end{aligned}$$

Taking into account that a similar inequality can be obtained for $\mu(\alpha_i + x\alpha_i^{1/2})$ we have Lemma 1.5.

1.4 The total number of particles visiting the unit ball

Since $\{s(t),\ t \geq 0\}$ is a stationary, ergodic process with $\mathbf{E}s(t) = \lambda$, we have

THEOREM 1.8

$$\lim_{T\to\infty} \frac{D(T)}{T} = \lambda \qquad a.s.$$

Our main goal in this Section is to prove that $D(T)$ also satisfies the central limit theorem. In fact we have

THEOREM 1.9

$$\lim_{T\to\infty} \mathbf{P}\left\{\frac{D(T)-\lambda T}{b(T)} < x\right\} = \Phi(x)$$

where

$$b(T) = \begin{cases} C_d T^{1/2} & \textit{if} \quad d \geq 3, \\ C_2 (T\log T)^{1/2} & \textit{if} \quad d = 2, \\ C_1 T^{3/4} & \textit{if} \quad d = 1. \end{cases}$$

At first we introduce a few notations and present a few lemmas. Let $W(t) \in \mathbb{R}^d$ be a Wiener process and

$$\begin{aligned} U &= |\{t:\ W(t) \in C(r_0 e_1, r_0)\}|, \\ U^+(T) &= |\{t:\ W(t) \in C(r_0 e_1, r_0),\ t \geq T\}|, \\ U^-(T) &= |\{t:\ W(t) \in C(r_0 e_1, r_0),\ t < T\}| \end{aligned}$$

where $e_1 = (1, 0, 0, \ldots, 0) \in \mathbb{R}^d$ and r_0 is the radius of C.

LEMMA 1.6 *Let $d \geq 3$. Then we have*

$$\mathbf{E}U = \frac{1}{2\pi^{d/2}}\Gamma\left(\frac{d-2}{2}\right)\int_C \frac{dx}{\|x - r_0 e_1\|^{d-2}}, \tag{1.4}$$

$$\mathbf{E}U^+(T) = \frac{2}{(2\pi)^{d/2}}\frac{1}{d-2}\frac{1+o(1)}{T^{(d-2)/2}} \quad \textit{as} \quad T \to \infty \tag{1.5}$$

and

$$\mathbf{E}U^2 \leq K < \infty \tag{1.6}$$

where $K = K(d)$ is a positive constant.

Note that if $d \geq 3$ then

$$\int_C \frac{dx}{\|x - r_0 e_1\|^{d-2}} < \infty.$$

Proof. Clearly

$$
\begin{aligned}
\mathbf{E}U &= \int_0^\infty \int_C (2\pi t)^{-d/2} \exp\left(-\frac{1}{2t}\|x - r_0 e_1\|^2\right) dx dt = \\
&= \int_C \int_0^\infty (2\pi t)^{-d/2} \exp\left(-\frac{1}{2t}\|x - r_0 e_1\|^2\right) dt dx = \\
&= \frac{2}{(2\pi)^{d/2}} \int_C \int_0^\infty \frac{u^{d-3}}{\|x - r_0 e_1\|^{d-2}} \exp\left(-\frac{u^2}{2}\right) du dx = \\
&= \frac{2}{(2\pi)^{d/2}} \int_C \frac{1}{\|x - r_0 e_1\|^{d-2}} \int_0^\infty u^{d-3} \exp\left(-\frac{u^2}{2}\right) du dx = \\
&= \frac{1}{2\pi^{d/2}} \Gamma\left(\frac{d-2}{2}\right) \int_C \frac{dx}{\|x - r_0 e_1\|^{d-2}}.
\end{aligned}
$$

Hence we have (1.4).

Similarly

$$
\begin{aligned}
\mathbf{E}U^+(T) &= \int_C \int_T^\infty (2\pi t)^{-d/2} \exp\left(-\frac{1}{2t}\|x - r_0 e_1\|^2\right) dt dx = \\
&= \int_C \int_0^{T^{-1/2}\|x - r_0 e_1\|} \frac{2}{(2\pi)^{d/2}} \frac{u^{d-3}}{\|x - r_0 e_1\|^{d-2}} \exp\left(-\frac{u^2}{2}\right) du dx = \\
&= \frac{2}{(2\pi)^{d/2}} \int_C \frac{1}{\|x - r_0 e_1\|^{d-2}} \int_0^{T^{-1/2}\|x - r_0 e_1\|} (1 + o(1)) u^{d-3} du dx = \\
&= \frac{2}{(2\pi)^{d/2}} \frac{1}{d-2} \frac{1 + o(1)}{T^{(d-2)/2}}.
\end{aligned}
$$

Hence we have (1.5).

For any $0 < t_1 < t_2 < \infty$ we have

$$
\mathbf{P}\{W(t_1) \in C(r), W(t_2) \in C(r)\} \le t_1^{-d/2} (t_2 - t_1)^{-d/2}
$$

Hence

$$
\begin{aligned}
\mathbf{E}U^2 \le 1 &+ \int_1^\infty t_1^{-d/2} dt_1 + \int_1^\infty t_2^{-d/2} dt_2 + \\
&+ \int_1^\infty \int_{t_1+1}^\infty t_1^{-d/2} (t_2 - t_1)^{-d/2} dt_2 dt_1 < \infty
\end{aligned}
$$

which implies (1.6) and we have the Lemma. An other form of the value of $\mathbf{E}U$ will be given in Lemma 1.12.

Let

$$
U_i = |\{t : P_i(t) \in C\}|
$$

and

$$
U^+(i, t) = |\{s : s \ge t, P_i(s)\}|.
$$

Further let $i_1, i_2, \dots, i_{N(t)}$ be the sequence of those i's for which $P_i(\cdot)$ visits C up to time t ordered in the order of their first visits. Then

$$D(t) = \sum_{j=1}^{N(t)} U_{i_j} - \sum_{j=1}^{N(t)} U^+(i_j, t). \tag{1.7}$$

LEMMA 1.7 *Let $d \geq 3$. Then we have*

$$\lim_{t\to\infty} \frac{1}{t} \sum_{j=1}^{N(t)} U_{i_j} = C_2(d) \qquad a.s. \tag{1.8}$$

where

$$C_2(d) = \frac{1}{2\pi^{d/2}} \Gamma\left(\frac{d-2}{2}\right) \int_C \frac{dx}{\|x - r_0 e_1\|^{d-2}} \lim_{t\to\infty} \frac{\mu(t)}{t}.$$

Further

$$\lim_{t\to\infty} \mathbf{P} \left\{ \frac{\sum_{j=1}^{N(t)} U_{i_j} - \mathbf{E}N(t)\mathbf{E}U}{C_3(d)t^{1/2}} < x \right\} = \Phi(x) \tag{1.9}$$

and

$$\limsup_{t\to\infty} \frac{\sum_{j=1}^{N(t)} U_{i_j} - \mathbf{E}N(t)\mathbf{E}U}{(2C_3(d)t \log\log t)^{1/2}} = 1 \qquad a.s. \tag{1.10}$$

where

$$C_3(d) = (C_2(d) \operatorname{Var} U)^{1/2}.$$

Proof. Clearly $U_{i_1}, U_{i_2}, \dots$ are i.i.d.r.v.'s independent from $N(t)$. Hence we have (1.8), (1.9) and (1.10) by (1.4) and Theorem 1.7.

LEMMA 1.8 ([44] Lemma 17.1) *Let $0 < a < b < c < \infty$. Then*

$$\mathbf{P}\{\inf\{s:\ s>0,\ \|W(t+s)\| = a\} < \inf\{s:\ s>0,\ \|W(t+s)\| = c\} \mid \|W(t)\| = b\} =$$

$$= \begin{cases} \dfrac{c-b}{c-a} & \text{if} \quad d = 1, \\[2ex] \dfrac{\log c - \log b}{\log c - \log a} & \text{if} \quad d = 2, \\[2ex] \dfrac{c^{2-d} - b^{2-d}}{c^{2-d} - a^{2-d}} & \text{if} \quad d \geq 3. \end{cases}$$

LEMMA 1.9 *Let $d \geq 3$ and*

$$W_x(t) = W(t) + x \qquad (x \in \mathbb{R}^d, x \notin C).$$

Then

$$\mathbf{P}\{\exists t: \ W_x(t) \in C\} = \left(\frac{r_0}{\|x\|}\right)^{d-2}.$$

Proof. Applying Lemma 1.8 with $a = r_0$, $b = \|x\|$, $c = \infty$, we get Lemma 1.9.

LEMMA 1.10 *Let $d \geq 3$. Then*

$$\mathbf{P}\{\exists t: \ t \geq T, \ W(t) \in C\} = \frac{C_1(d) + o(1)}{T^{(d-2)/2}}$$

where

$$C_1(d) = \frac{r_0^{d-2}}{2^{d/2}} \frac{d}{\Gamma(d/2+1)} \int_{r_0}^{\infty} u e^{-u^2/2} du = \frac{d r_0^{d-2}}{2^{d/2}\Gamma(d/2+1)} \exp\left(-\frac{r_0^2}{2}\right).$$

Proof. By Lemma 1.9

$$\mathbf{P}\{\exists t: \ t \geq T, \ W(t) \in C \mid W(T)\} = \begin{cases} \left(\dfrac{r_0}{\|W(T)\|}\right)^{d-2} & \text{if} \quad \|W(T)\| > r_0, \\ 1 & \text{if} \quad \|W(T)\| \leq r_0. \end{cases}$$

Taking into account that the surface of a sphere of radius r is

$$\frac{d\pi^{d/2} r^{d-1}}{\Gamma(d/2+1)}$$

we have

$$\begin{aligned} &\mathbf{P}\{\exists t: \ t \geq T, \ W(t) \in C\} = \\ &= \mathbf{P}\{\|W(T)\| \leq r_0\} + \mathbf{E}\left(\left(\frac{r_0}{\|W(T)\|}\right)^{d-2} \Bigg| \ \|W(T)\| > r_0\right) \mathbf{P}\{\|W(T)\| > r_0\} = \\ &= \frac{1+o(1)}{(2\pi T)^{d/2}} + \frac{1+o(1)}{T^{(d-2)/2}} \frac{r_0^{d-2}}{(2\pi)^{d/2}} \int_{\mathbb{R}^d - C} \frac{1}{\|x\|^{d-2}} e^{-\|x\|^2/2} dx = \\ &= \frac{1+o(1)}{(2\pi T)^{d/2}} + \frac{1+o(1)}{T^{(d-2)/2}} \frac{r_0^{d-2}}{(2\pi)^{d/2}} \frac{\pi^{d/2} d}{\Gamma(d/2+1)} \int_{r_0}^{\infty} u e^{-u^2/2} du \end{aligned}$$

and we have Lemma 1.10.

LEMMA 1.11 *There exists an $L>0$ such that*

$$\lim_{t\to\infty} t^{-1/2}\sum_{j=1}^{N(t)} U^+(i_j,t) = \begin{cases} L & \text{if} \quad d=3, \\ 0 & \text{if} \quad d\geq 4 \end{cases}$$

almost surely.

Proof. Let

$$\tau_j = \min\{s:\ P_{i_j}(s)\in C\} \qquad (j=1,2,\ldots),$$
$$I_j = \begin{cases} 1 & \text{if} \quad \{\exists s:\ s\geq t,\ P_{i_j}(s)\in C\}, \\ 0 & \text{otherwise.} \end{cases}$$

Then by Lemma 1.10

$$\mathbf{P}\{I_j=1\} = \mathbf{E}\mathbf{P}\{I_j=1\mid \tau_j\} = \mathbf{E}\frac{C_1(d)+o(1)}{(t-\tau_j)^{(d-2)/2}}$$

if $d\geq 3$. As a simple consequence of Lemma 1.5 we get

$$\mathbf{E}(t-\tau_j)^{(2-d)/2} = \begin{cases} \left(t-\dfrac{(1+o(1))j}{\lambda(6\pi^2)^{1/3}}\right)^{-1/2} & \text{if} \quad d=3, \\ \left(t-\dfrac{(1+o(1))j}{\lambda C_d}\right)^{(2-d)/2} & \text{if} \quad d\geq 4. \end{cases}$$

Consider those r.v.'s $I_jU^+(i_j,t)$ for which $I_j=1$. Clearly these r.v.'s are i.i.d. with

$$\mathbf{P}\{I_jU^+(i_j,t)<x\mid I_j=1\} = \mathbf{P}\{U<x\}.$$

Hence

$$\mathbf{E}U^+(i_j,t) = \begin{cases} (C_1(3)+o(1))\left(t-\dfrac{(1+o(1))j}{\lambda(6\pi^2)^{1/3}}\right)^{-1/2}\mathbf{E}U & \text{if} \quad d=3, \\ (C_1(d)+o(1))\left(t-\dfrac{(1+o(1))j}{\lambda C_d}\right)^{(2-d)/2}\mathbf{E}U & \text{if} \quad d\geq 4 \end{cases}$$

and

$$\mathbf{E}\sum_{j=1}^{N(t)} U^+(i_j,t) = \begin{cases} (1+o(1))Lt^{1/2} & \text{if} \quad d=3, \\ o(t^{1/2}) & \text{if} \quad d\geq 4. \end{cases}$$

Hence we have Lemma 1.11.

LEMMA 1.12

$$\mathbf{E}U = \begin{cases} (6\pi^2)^{-1/3} & \text{if} \quad d=3, \\ C_d^{-1} & \text{if} \quad d\geq 4 \end{cases} \tag{1.11}$$

and

$$L = \left(\frac{3^2\cdot 2^5}{\pi^5}\right)^{1/6} \quad \text{if} \quad d=3 \tag{1.12}$$

where C_d is the constant of Theorem 1.3 and L is defined in Lemma 1.11.

Proof. Since $\mathbf{E}D(t) = \lambda t$, by (1.7) we have

$$\lambda t = \mathbf{E}N(t)\mathbf{E}U - \mathbf{E}\sum_{j=1}^{N(t)} U^+(i_j,t). \tag{1.13}$$

By Lemma 1.11 and (1.2)

$$\lambda = \mathbf{E}U \lim_{t\to\infty} \frac{\mu(t)}{t} \quad \text{if} \quad d \geq 3.$$

Hence we have (1.11).

In case $d = 3$ by (1.2), (1.11) and (1.13)

$$\lambda t = \lambda\left((6\pi^2)^{1/3}t + \left(\frac{3^4\cdot 2^7}{\pi}\right)^{1/6} t^{1/2}\right)(6\pi^2)^{-1/3} - \mathbf{E}\sum_{j=1}^{N(t)} U^+(i_j,t)$$

which, in turn, implies (1.12).

Proof of Theorem 1.9 in case $d \geq 3$. By (1.7) we have

$$D(t) - \lambda t = \sum_{j=1}^{N(t)} U_{i_j} - \mathbf{E}N(t)\mathbf{E}U - \left(\sum_{j=1}^{N(t)} U^+(i_j,t) - K_d(t)\right) - \\ - (\lambda t - \mathbf{E}N(t)\mathbf{E}U + K_d(t))$$

where

$$K_d(t) = \begin{cases} L & \text{if} \quad d = 3, \\ 0 & \text{if} \quad d \geq 4. \end{cases}$$

Observe that by Lemmas 1.7 and 1.11

$$\lim_{t\to\infty} \mathbf{P}\left\{\frac{\sum_{j=1}^{N(t)} U_{i_j} - \mathbf{E}N(t)\mathbf{E}U}{C_3(d)t^{1/2}} < x\right\} = \Phi(t)$$

and

$$\lim_{t\to\infty} \frac{\sum_{j=1}^{N(t)} U^+(i_j,t) - K_d(t)}{t^{1/2}} = 0 \qquad \text{a.s.}$$

Since by (1.13) and Lemma 1.12

$$\lim_{t\to\infty} \frac{\lambda t - \mathbf{E}N(t)\mathbf{E}U + K_d(t)}{t^{1/2}} = 0,$$

we have Theorem 1.9 for $d \geq 3$.

In order to prove the Theorem in cases $d = 1$ and $d = 2$ we present two further lemmas.

LEMMA 1.13 *Let $d = 2$. Then we have*

$$\mathbf{E}U^-(T) \sim \frac{1}{2\pi}\log T \tag{1.14}$$

and

$$\operatorname{Var} U^-(T) = (K + o(1))(\log T)^2. \tag{1.15}$$

Proof. Clearly

$$\begin{aligned}\mathbf{E}U^-(T) &= \int_0^T \int_C (2\pi t)^{-1} \exp\left(-\frac{1}{2t}\|x - r_0 e_1\|^2\right) dx dt = \\ &= (2\pi)^{-1/2} \int_C \sqrt{\frac{2}{\pi}} \int_{\|x - r_0 e_1\| T^{-1/2}}^{\infty} u^{-1} \exp\left(-\frac{u^2}{2}\right) du dx \sim \\ &\sim \frac{1}{2\pi}\log T.\end{aligned}$$

Hence we have (1.14). (1.15) can be obtained combining the methods of proofs of (1.14) and (1.6).

This result easily implies Theorem 1.9 in case $d = 2$.

LEMMA 1.14 *Let $d = 1$. Then we have*

$$\mathbf{E}U^-(T) \sim \left(\frac{2T}{\pi}\right)^{1/2},$$

$$\operatorname{Var} U^-(T) \sim (K + o(1))T.$$

Proof is the same as that of Lemma 1.11 and we obtain Theorem 1.9 in case $d = 1$.

By (1.7), (1.10) and Lemma 1.11 we also have

THEOREM 1.10 *Let $d \geq 3$. Then*

$$\limsup_{t\to\infty} \frac{D(t) - \lambda t}{(2C_3(d)t\log\log t)^{1/2}} = 1 \qquad a.s.$$

1.5 Death in C

Let us reformulate the language of Section 1.3 saying that a particle dies when it enters to C at first. Then clearly $N(t)$ is the number of died particles at time t. Theorem 1.3 suggests that around the origin there exists a large empty ball if t is large and $d = 1$ or 2. (On empty ball we mean a ball not containing any living particle.) Now we prove that it is really so.

In order to formulate our results introduce the following notations. Let

$$\begin{aligned}T_t &= \min\{s: \ s > 0, \ N(t+s) - N(t) \geq 1\}, \\ \hat{s}(A, t) &= \#\{i: \ P_i(t) \in A, \ P_i(s) \notin C \text{ for all } s < t\}, \\ X_t &= \min\{R: \ \hat{s}(C(R), t) \geq 1\}.\end{aligned}$$

Clearly T_t is the length of the longest time–interval begining at t in which no living particle enters to C, $\hat{s}(\mathcal{A},t)$ is the number of living particles located in $\mathcal{A}$ at t and X_t is the radius of the largest ball around the origin not containing any living particle at time t.

Now we formulate our

THEOREM 1.11

$$\lim_{t\to\infty} \mathbf{P}\{T_t > x g_d(t)\} = \exp(-\lambda c_d x) \qquad (x > 0) \tag{1.16}$$

where

$$g_d(t) = \begin{cases} t^{1/2} & \textit{if} \quad d = 1, \\ \log t & \textit{if} \quad d = 2, \\ 1 & \textit{if} \quad d \geq 3, \end{cases}$$

$$c_d = \begin{cases} \left(\dfrac{2}{\pi}\right)^{1/2} & \textit{if} \quad d = 1, \\ 2\pi & \textit{if} \quad d = 2, \\ (6\pi^2)^{1/3} & \textit{if} \quad d = 3, \\ C_d & \textit{if} \quad d \geq 4 \end{cases}$$

and C_d is the constant of (1.2).

Proof. Observe that by Theorems 1.3 and 1.5

$$\mathbf{P}\{T_t > z\} = \mathbf{P}\{N(t+z) - N(t) = 0\} = \exp(-(\mu(t+z) - \mu(t)))$$

and

$$\lim_{t\to\infty} (\mu(t + x g_d(t)) - \mu(t)) = \lambda c_d x. \tag{1.17}$$

Hence we have Theorem 1.11.

THEOREM 1.12 *Let $d = 1$. Then*

$$\lim_{t\to\infty} \mathbf{P}\{X_t > x t^{1/4}\} = \exp\left(-\lambda \left(\frac{2}{\pi}\right)^{1/2} x^2\right) \qquad (x > 0). \tag{1.18}$$

Proof. For any $z > 0,\ y > 0,\ t > 0,\ 0 < \alpha < 1/2$ let

$$\begin{aligned} f = f(z,y,t,\alpha) &= \mathbf{P}\{\max_{0\leq s\leq t} W(s) < z,\ W(t) > z - yt^\alpha\}, \\ A &= \{\max_{0\leq s\leq t} W(s) < z\}, \\ B &= \{z - yt^\alpha < W(t) < z\}, \\ g = g(z,y,t,\alpha) &= \mathbf{P}\{\overline{A}B\}. \end{aligned}$$

Then

$$f = \mathbf{P}(AB) = \mathbf{P}(B) - \mathbf{P}(\overline{A}B),$$
$$\mathbf{P}(B) = \Phi(zt^{-1/2}) - \Phi(zt^{-1/2} - yt^{\alpha-1/2})$$

and by the reflection principle (cf. [44] p. 15)

$$\mathbf{P}(\overline{A}B) = \mathbf{P}\{z < W(t) < z + yt^{\alpha}\} = \\ = \Phi(zt^{-1/2} + yt^{\alpha-1/2}) - \Phi(zt^{-1/2}).$$

Hence

$$f = \Phi(zt^{-1/2}) - \Phi(zt^{-1/2} - yt^{\alpha-1/2}) - (\Phi(zt^{-1/2} + yt^{\alpha-1/2}) - \Phi(zt^{-1/2})).$$

Let $z = \varsigma t^{1/2}$. Then

$$f(\varsigma t^{1/2}, y, t, \alpha) = \Phi(\varsigma) - \Phi(\varsigma - yt^{\alpha-1/2}) - (\Phi(\varsigma + yt^{\alpha-1/2}) - \Phi(\varsigma)) \sim \\ \sim -(yt^{\alpha-1/2})^2\Phi''(\varsigma) = y^2t^{2\alpha-1}(2\pi)^{-1/2}\varsigma e^{-\varsigma^2/2} = \\ = (2\pi)^{-1/2}y^2t^{2\alpha-3/2}z\exp\left(-\frac{z^2}{2t}\right).$$

Not taking into account those elements of π which are lying in $[-1/2, 1/2]$ let

$$\pi_0 = \{\ldots < z_{-2} < z_{-1} < -1/2 < 1/2 < z_1 < z_2 < \ldots\}.$$

Then

$$\mathbf{P}\{X_t > xt^{1/4} \mid \pi_0\} = \prod_{i=-\infty}^{+\infty}(1 - f(|z_i|, x, t, 1/4)).$$

It is easy to see (cf. Lemma 1.5) that

$$\mathbf{P}\left\{\left|z_i - \frac{i}{\lambda}\right| \geq y\left(\frac{i}{\lambda}\right)^{1/2}\right\} \leq \exp(-Cy^2)$$

and by a simple calculation we obtain

$$\sum_{i=-\infty}^{+\infty} f(|z_i|, x, t, 1/4) \sim (2\pi)^{1/2}\frac{x^2}{t}\sum_{i=-\infty}^{+\infty}|z_i|\exp\left(-\frac{z_i^2}{2t}\right) \sim \\ \sim 2(2\pi)^{-1/2}\frac{x^2}{\lambda t}\sum_{i=1}^{\infty} i\exp\left(-\frac{i^2}{2\lambda^2 t}\right) = \\ = \left(\frac{2}{\pi}\right)^{1/2}x^2\lambda\sum_{i=1}^{\infty}\frac{i}{\lambda t^{1/2}}\exp\left(-\frac{1}{2}\left(\frac{i}{\lambda t^{1/2}}\right)^2\right)\frac{1}{\lambda t^{1/2}} \sim \\ \sim \left(\frac{2}{\pi}\right)^{1/2}x^2\lambda\int_0^{\infty} u\exp\left(-\frac{u^2}{2}\right)du = \left(\frac{2}{\pi}\right)^{1/2}x^2\lambda.$$

Consequently

$$\lim_{t\to\infty} \mathbf{EP}\{X_t < xt^{1/4} \mid \pi_0\} = \exp\left(-\lambda\left(\frac{2}{\pi}\right)^{1/2} x^2\right) \qquad \text{a.s.}$$

and we have (1.18).

In case $d = 2$ we do not have such an exact result like Theorem 1.12 in case $d = 1$. We only prove

THEOREM 1.13 *Let $d = 2$. Then for any $0 < \delta < 1/2$ and $0 < \alpha < 1$ we have*

$$\mathbf{P}\{X_t > (\log t)^{1/2-\delta}\} > \alpha$$

if t is big enough.

At first we recall

LEMMA 1.15 ([45] Lemma 1) *Let $u \in \mathbb{R}^2$ satisfying the inequality*

$$er_0 \leq \|u\| \leq T^{1/2-\delta}.$$

Then for any $0 < \varepsilon < \delta < 1/2$ there exists a $T_0 = T_0(\varepsilon, \delta) > 0$ such that

$$1 - (1+\varepsilon)\frac{\log(\|u\|r_0^{-1})}{\log(T^{1/2-\varepsilon}r_0^{-1})} \leq \mathbf{P}\{\exists t:\ 0 \leq t \leq T,\ W(t) \in C(u, r_0)\} \leq$$

$$\leq 1 - (1-\varepsilon)\frac{\log(\|u\|r_0^{-1})}{\log(T^{1/2+\varepsilon}r_0^{-1})}$$

if $T \geq T_0$.

Proof of Theorem 1.12. Apply Lemma 1.15 with

$$T = x\log t, \quad \|u\| = (\log t)^{1/2-\delta}, \quad x > 0.$$

Then

$$\mathbf{P}\{\exists s:\ t \leq s \leq t + x\log t,\ P_i(s) \in C(0, r_0) \mid \|P_i(t)\| \leq (\log t)^{1/2-\delta}\} \geq$$

$$\geq 1 - (1+\varepsilon)\frac{\log((\log t)^{1/2-\delta}r_0^{-1})}{\log(x(\log t)^{1/2+\varepsilon}r_0^{-1})} \geq 2\delta - \varepsilon$$

for any $x > 0$ if t is big enough.

Let

$$A_t = \{\exists i:\ \|P_i(t)\| \leq (\log t)^{1/2-\delta},\ P_i(s) \notin C,\ 0 \leq s \leq t\} = \{X_t \leq (\log t)^{1/2-\delta}\}.$$

Then

$$\mathbf{P}\{T_t < x\log t \mid A_t\} \geq 2\delta - \varepsilon$$

and

$$\mathbf{P}\{A_t\} \leq \frac{\mathbf{P}\{T_t < x\log t, A_t\}}{2\delta - \varepsilon} \leq \frac{\mathbf{P}\{T_t < x\log t\}}{2\delta - \varepsilon}.$$

Choosing x small enough, by Theorem 1.11 we have Theorem 1.13.

1.6 Charged particles

Consider the process

$$P(t) = \{P_1(t), P_2(t), \ldots\}$$

defined in Section 1.2 and let $\sigma_1, \sigma_2, \ldots$ be a sequence of i.i.d.r.v.'s independent from $P(t)$ with

$$\mathbf{P}\{\sigma_1 = 1\} = \mathbf{P}\{\sigma_1 = -1\} = 1/2.$$

Introduce the following notations. For any Borel set $A \subset \mathbb{R}^d$ let

$$\begin{aligned}
P^+(t) &= \{\delta_1 P_1(t), \delta_2 P_2(t), \ldots\} \quad \text{where} \quad \delta_i = (\sigma_i + 1)/2, \\
P^-(t) &= \{\Delta_1 P_1(t), \Delta_2 P_2(t), \ldots\} \quad \text{where} \quad \Delta_i = (-\sigma_i + 1)/2, \\
s^+(A,t) &= \#\{i : P_i(t) \in A, \ \sigma_i = +1\}, \\
s^-(A,t) &= \#\{i : P_i(t) \in A, \ \sigma_i = -1\}, \\
s^{+,-}(A,t) &= s^+(A,t) - s^-(A,t), \\
s^+(t) &= s^+(C,t), \\
s^-(t) &= s^-(C,t), \\
s^{+,-}(t) &= s^+(t) - s^-(t), \\
S^{+,-}(T) &= \sup_{0 \le t \le T} s^{+,-}(t), \\
N^+(t) &= \#\{i : \sigma_i = 1, \ \exists 0 \le s \le t \text{ such that } P_i(s) \in C\}, \\
N^-(t) &= \#\{i : \sigma_i = -1, \ \exists 0 \le s \le t \text{ such that } P_i(s) \in C\}, \\
N^{+,-}(t) &= N^+(t) - N^-(t), \\
D^{+,-}(T) &= \int_0^T s^{+,-}(t)dt.
\end{aligned}$$

If one were to think of the random signs as signed particles moving throughout space according to independent Wiener processes, then $P^+(t)$ resp. $P^-(t)$ would be the Poisson random field generated by the particles of charge $+1$ resp. -1, $s^{+,-}(t)$ would describe the charge in the set C at time t. Similarly $S^{+,-}(T)$ is the largest charge in C up to time T, $N^{+,-}(t)$ is the sum of the charges of the died particles at t (the word "died" is used like in Section 1.5). The meaning of $D^{+,-}(T)$ is clear as it is defined.

Observe that $P^+(t)$ and $P^-(t)$ for any $t \ge 0$ by Lemma 1.2 and Theorem 1.1. are independent Poisson random fields of parameters $\lambda/2$. Having this observation without any new idea one can prove the following:

THEOREM 1.14 (i) *For any fixed t $s^{+,-}(A,t)$ is a process of independent increments i.e. $s^{+,-}(A,t)$ and $s^{+,-}(B,t)$ are independent r.v.'s if $AB = \emptyset$ and*

$$\mathbf{P}\{s^{+,-}(t) = k\} = \left(\frac{\lambda}{2}\right)^{|k|} e^{-\lambda} \sum_{n=0}^{\infty} \frac{\lambda^{2n}}{2^{2n} n!(|k|+n)!} \qquad (k = 0, \pm 1, \pm 2, \ldots).$$

(ii) $N^+(t)$ *and* $N^-(t)$ $(t \geq 0)$ *are independent Poisson processes* $\{N^{+,-}(t),\ t \geq 0\}$ *is a process of independent increments with*

$$\lim_{t\to\infty} \mathbf{P}\left\{\frac{N^{+,-}(t)}{(\mathbf{E}N(t))^{1/2}} < x\right\} = \Phi(x) \qquad (-\infty < x < \infty)$$

where $\mathbf{E}N(t)$ *is given by* (1.2).

(iii) $D^+(t)$ *and* $D^-(t)$ *are independent processes and*

$$\lim_{t\to\infty} \mathbf{P}\left\{\frac{D^{+,-}(t)}{b(t)} < x\right\} = \Phi(x) \qquad (-\infty < x < \infty)$$

where $b(t)$ *is given in Theorem* 1.9.

Note that Theorems 1.6 and 1.7 can be also extended for $N^{+,-}(t)$ without any difficulty.

1.7 The independence of the increments

In Section 1.3 we have seen that for any fixed $t \geq 0$ the process $s(A,t)$ is a Poisson point process. Consequently it is a process of independent increments i.e. $s(A,t)$ and $s(B,t)$ are independent r.v.'s if $AB = \emptyset$. However it is easy to see that $N(A,t)$ and $D(A,t)$ are processes of non-independent increments. In this Section we show that $N(A,t)$ and $N(B,t)$ resp. $D(A,t)$ and $D(B,t)$ are "almost" independent if A and B are far away from each other. In fact we have

THEOREM 1.15 *Let* $f_i(t)$ $(i = 1,2,3;\ t \geq 0)$ *be positive valued functions with*

$$\lim_{t\to\infty} t^{-1/2} f_1(t) = \infty,$$

$$\lim_{t\to\infty} \frac{\log(t^{1/2}(f_2(t))^{-1})}{\log t} = 0 \quad \textit{and} \quad f_2(t) \leq t^{1/2},$$

$$\lim_{t\to\infty} f_3(t) = \infty.$$

Then

$$\lim_{t\to\infty} \mathbf{P}\left\{\frac{N(t) - \mathbf{E}N(t)}{(\mathbf{E}N(t))^{1/2}} < x,\ \frac{N(C(f),t) - \mathbf{E}N(C(f),t)}{(\mathbf{E}N(C(f),t))^{1/2}} < y\right\} = \Phi(x)\Phi(y) \quad (1.19)$$

and

$$\lim_{t\to\infty} \mathbf{P}\left\{\frac{D(t) - \lambda t}{b(t)} < x,\ \frac{D(C(f),t) - \lambda t}{b(t)} < y\right\} = \Phi(x)\Phi(y) \quad (1.20)$$

where

$$C(f) = C(f(t)) = C(f(t), r_0) = \{x :\ x \in \mathbb{R}^d,\ \|x - f(t)\| \leq r_0\},$$

$r_0^{-d} = \omega_d$ *is defined in the proof of Theorem* 1.3, $b(t)$ *is defined in Theorem* 1.9 *and*

$$f = f_1 \quad \textit{if} \quad d = 1, \qquad f \geq f_2 \quad \textit{if} \quad d = 2, \qquad f = f_3 \quad \textit{if} \quad d \geq 3.$$

Note that

$$f_2(t) = t^{1/2}(\log t)^{-\alpha}$$

satisfies the above condition for any $\alpha > 0$.

Before the proof of Theorem 1.15 we present our

LEMMA 1.16 *Let $\{W(t),\ t \geq 0\}$ be a Wiener process in $\mathbb{R}^d$. Then*

$$\lim_{t\to\infty} \mathbf{P}\{\exists s:\ 0 \leq s \leq t,\ W(s) \in C(f_d)\} = 0$$

where f_d satisfies the conditions of Theorem 1.15.

Proof. The proof is trivial in cases $d = 1$ and $d \geq 3$. In case $d = 2$ observe that for any $f_2(\cdot)$ satisfying the condition of Theorem 1.15 there exists a function $g(\cdot)$ such that

$$g(t) \nearrow \infty \quad \text{and} \quad \frac{\log(t^{1/2}g(t)(f_2(t))^{-1})}{\log t} \to 0 \quad \text{as} \quad t \to \infty.$$

Let

$$\tau(x,\rho) = \inf\{s:\ s \geq 0,\ W(s) \in \partial C(x,\rho)\} \quad (x \in \mathbb{R}^2,\ \rho > 0).$$

Then

$$\begin{aligned}
&\mathbf{P}\{\exists s:\ 0 \leq s \leq t,\ W(s) \in C(f_2)\} \leq \\
&\leq \mathbf{P}\{\tau(f_2, r_0) \leq \tau(f_2, t^{1/2}g(t)) \text{ or } \tau(f_2, t^{1/2}g(t)) \leq t\} \leq \\
&\leq \mathbf{P}\{\tau(f_2, r_0) \leq \tau(f_2, t^{1/2}g(t))\} + \mathbf{P}\{\tau(f_2, t^{1/2}g(t)) \leq t\}.
\end{aligned}$$

By Lemma 1.8 we have

$$\mathbf{P}\{\tau(f_2, r_0) \leq \tau(f_2, t^{1/2}g(t))\} = \frac{\log t^{1/2}g(t) - \log f_2(t)}{\log t^{1/2}g(t) - \log r_0} \to 0.$$

Since

$$\lim_{t\to 0} \mathbf{P}\{\tau(f_2, t^{1/2}g(t)) \leq t\} = 0$$

we have Lemma 1.16.

Proof of Theorem 1.15. Let

$$\begin{aligned}
N_1(t) &= \#\{i:\ \exists 0 \leq s \leq t:\ P_i(s) \in C,\ \forall 0 \leq s \leq t:\ P_i(s) \notin C(f)\},\\
N_2(t) &= \#\{i:\ \exists 0 \leq s \leq t:\ P_i(s) \in C(f),\ \forall 0 \leq s \leq t:\ P_i(s) \notin C\},\\
N_{1,2}(t) &= \#\{i:\ \exists 0 \leq s_1 < s_2 \leq t:\ P_i(s_1) \in C,\ P_i(s_2) \in C(f),\\
&\qquad \forall 0 \leq s \leq s_1:\ P_i(s) \notin C(f)\},\\
N_{2,1}(t) &= \#\{i:\ \exists 0 \leq s_1 < s_2 \leq t:\ P_i(s_1) \in C(f),\ P_i(s_2) \in C,\\
&\qquad \forall 0 \leq s \leq s_1:\ P_i(s) \notin C\}
\end{aligned}$$

i.e.

$N_1(t)$ is the number of those particles which visit C but do not visit $C(f)$ up to t,

$N_2(t)$ is the number of those particles which visit $C(f)$ but do not visit C up to t,

$N_{1,2}(t)$ is the number of those particles which visit both C and $C(f)$ up to t but visit C before $C(f)$,

$N_{2,1}(t)$ is the number of those particles which visit both C and $C(f)$ up to t but visit $C(f)$ before C.

Observe that

$$N(t) = N_1(t) + N_{1,2}(t) + N_{2,1}(t), \tag{1.21}$$

and

$$N(C(f),t) = N_2(t) + N_{1,2}(t) + N_{2,1}(t) \tag{1.22}$$

where by Lemma 1.1 $N_1(t)$, $N_2(t)$, $N_{1,2}(t)$, $N_{2,1}(t)$ are independent Poisson r.v.'s (cf. also the proof of Theorem 1.3). Note also that by Lemma 1.16 we have

$$\mathbf{E}N_{1,2}(t) = o(\mathbf{E}N(t)) = \mathbf{E}N_{21}(t). \tag{1.23}$$

Consider

$$\begin{aligned}\frac{N(t)-\mathbf{E}N(t)}{(\mathbf{E}N(t))^{1/2}} = {} & \frac{N_1(t)-\mathbf{E}N_1(t)}{(\mathbf{E}N_1(t))^{1/2}}\left(\frac{\mathbf{E}N_1(t)}{\mathbf{E}N(t)}\right)^{1/2} + \\ & + \frac{N_{1,2}(t)-\mathbf{E}N_{1,2}(t)}{(\mathbf{E}N_{1,2}(t))^{1/2}}\left(\frac{\mathbf{E}N_{1,2}(t)}{\mathbf{E}N(t)}\right)^{1/2} + \\ & + \frac{N_{2,1}(t)-\mathbf{E}N_{2,1}(t)}{(\mathbf{E}N_{2,1}(t))^{1/2}}\left(\frac{\mathbf{E}N_{2,1}(t)}{\mathbf{E}N(t)}\right)^{1/2},\end{aligned}$$

similarly

$$\begin{aligned}\frac{N(C(f),t)-\mathbf{E}N(C(f),t)}{(\mathbf{E}N(C(f),t))^{1/2}} = {} & \frac{N_2(t)-\mathbf{E}N_2(t)}{(\mathbf{E}N_2(t))^{1/2}}\left(\frac{\mathbf{E}N_2(t)}{\mathbf{E}N(C(f),t)}\right)^{1/2} + \\ & + \frac{N_{1,2}(t)-\mathbf{E}N_{1,2}(t)}{(\mathbf{E}N_{1,2}(t))^{1/2}}\left(\frac{\mathbf{E}N_{1,2}(t)}{\mathbf{E}N(C(f),t)}\right)^{1/2} + \\ & + \frac{N_{2,1}(t)-\mathbf{E}N_{2,1}(t)}{(\mathbf{E}N_{2,1}(t))^{1/2}}\left(\frac{\mathbf{E}N_{2,1}(t)}{\mathbf{E}N(C(f),t)}\right)^{1/2}.\end{aligned}$$

Since by (1.23)

$$\frac{N_{1,2}(t)-\mathbf{E}N_{1,2}(t)}{(\mathbf{E}N_{1,2}(t))^{1/2}}\left(\frac{\mathbf{E}N_{1,2}(t)}{\mathbf{E}N(t)}\right)^{1/2} \to 0 \text{ in probability as } t \to \infty,$$

$$\frac{N_{2,1}(t) - \mathbf{E}N_{2,1}(t)}{(\mathbf{E}N_{2,1}(t))^{1/2}} \left(\frac{\mathbf{E}N_{2,1}(t)}{\mathbf{E}N(t)}\right)^{1/2} \to 0 \text{ in probability as } t \to \infty$$

and $N_1(t)$ and $N_2(t)$ are independent, Poisson r.v.'s, we have (1.19).

Repeating the above given proof without any new idea we get (1.20).

It is also worth-while to prove

LEMMA 1.17 *Let $d \geq 3$ and $u \in \mathbb{R}^d$ with $\|u\| = R$. Then for any $t \geq 0$*

$$\mathrm{Cov}(N(t), N(C(u, r_0), t)) \leq O(R^{2-d}). \tag{1.24}$$

Proof. It is known that

$$\mathbf{P}\{W(t) \in C(u, r_0) \text{ for some } t\} = \left(\frac{r_0}{R}\right)^{d-2}$$

if $d \geq 3$ (cf. [44], Lemma 22.16). Then by (1.21) and (1.22) we have

$$\mathrm{Cov}(N(t), N(C(u, r_0), t)) = \frac{2\mathrm{Var}N_{1,2}(t)}{\mathrm{Var}N(t) + 2\mathrm{Var}N_{1,2}(t)}. \tag{1.25}$$

By Theorem 1.3

$$\mathrm{Var}\, N(t) = \mathbf{E}N(t) = \lambda(C_d + o(1))t \tag{1.26}$$

and

$$\mathrm{Var}\, N_{1,2}(t) = \mathbf{E}N_{1,2}(t).$$

Observe that

$$\mathbf{E}(N_{1,2}(t) \mid N(s),\ s \leq t) \leq N(t)\mathbf{P}\{W(t) \in C(u, r_0) \text{ for some } t\} = N(t)\left(\frac{r_0}{R}\right)^{d-2}$$

Hence

$$\mathbf{E}N_{1,2}(t) \leq \lambda(C_d + o(1))t \left(\frac{r_0}{R}\right)^{d-2}$$

which implies

$$\mathrm{Var}\, N_{1,2}(t) \leq \lambda(C_d + o(1))t \left(\frac{r_0}{R}\right)^{d-2}. \tag{1.27}$$

Then by (1.25), (1.26) and (1.27) we have (1.24).

1.8 Brownian density process

As we have seen in Section 1.2, for any $t \geq 0$, the r.v. $s(t)$ is Poisson of parameter λ, but to describe the properties of the process $\{s(t),\ t \geq 0\}$ looks hard. However, it turns out that if $\lambda \to \infty$ then $\{s(t),\ t \geq 0\}$ converges in distribution to a Gaussian process $\Gamma(t)$, the so-called Brownian density process.

LEMMA 1.18

$$\mathbf{E}\left(\left(\frac{s(0)-\lambda}{\lambda^{1/2}}\right)\left(\frac{s(t)-\lambda}{\lambda^{1/2}}\right)\right) = (1+o(1))(2\pi t)^{-d/2}$$

as $t \to \infty$ where $o(1)$ does not depend on λ.

Proof. Let

$$b(t) = \#\{i : \; P_i(0) \in C, \; P_i(t) \in C\},$$
$$k(t) = \#\{i : \; P_i(0) \notin C, \; P_i(t) \in C\}.$$

Then

$$s(t) = b(t) + k(t)$$

and

$$\lambda = \mathbf{E}s(t) = \mathbf{E}(\mathbf{E}(b(t) \mid s(0))) + \mathbf{E}k(t).$$

Since

$$\mathbf{P}\{P_i(t) \in C \mid P_i(0) \in C\} = (1+o(1))(2\pi t)^{-d/2},$$

we have

$$\mathbf{E}(b(t) \mid s(0)) = (1+o(1))s(0)(2\pi t)^{-d/2},$$
$$\mathbf{E}b(t) = (1+o(1))\lambda(2\pi t)^{-d/2}$$

and

$$\mathbf{E}k(t) = \lambda(1-(1+o(1))(2\pi t)^{-d/2}).$$

Then

$$\mathbf{E}s(0)s(t) = \mathbf{E}(s(0)\mathbf{E}(s(t) \mid s(0)))$$

and

$$\mathbf{E}(s(t) \mid s(0)) = \mathbf{E}(b(t) \mid s(0)) + \mathbf{E}(k(t) \mid s(0)) =$$
$$= (1+o(1))s(0)(2\pi t)^{-d/2} + \lambda(1-(1+o(1))(2\pi t)^{-d/2}).$$

Hence

$$\mathbf{E}s(0)s(t) = (1+o(1))(2\pi t)^{-d/2}\mathbf{E}s^2(0) + \lambda^2(1-(1+o(1))(2\pi t)^{-d/2}) =$$
$$= \lambda^2 + (1+o(1))\lambda(2\pi t)^{-d/2}$$

which, in turn, implies Lemma 1.18.

LEMMA 1.19 *For any $0 \le t_1 < t_2 < \infty$ we have*

$$\lim_{\lambda\to\infty} \mathbf{P}\left\{\frac{s(t_1)-\lambda}{\lambda^{1/2}} < x\right\} = \Phi(x) \tag{1.28}$$

and

$$\lim_{\lambda\to\infty} \mathbf{P}\left\{\frac{s(t_1)-\lambda}{\lambda^{1/2}} < x,\ \frac{s(t_2)-\lambda}{\lambda^{1/2}} < y\right\} = \mathbf{P}\{N_1 < x,\ N_2 < y\} \tag{1.29}$$

where

$$N_1 \in \mathcal{N}(0,1), \qquad N_2 \in \mathcal{N}(0,1)$$

and

$$\mathbf{E}N_1N_2 = \operatorname{Cov}(s(0), s(t_2-t_1)).$$

Proof. (1.28) is trivial. In order to see (1.29) consider

$$Q = \mathbf{P}\left\{\frac{s(t)-\lambda}{\lambda^{1/2}} < x \;\middle|\; s(0) = [\lambda + y\lambda^{1/2}]\right\} =$$
$$= \mathbf{P}\left\{\frac{k(t)-\lambda}{\lambda^{1/2}} + \frac{b(t)}{\lambda^{1/2}} < x \;\middle|\; s(0) = [\lambda + y\lambda^{1/2}]\right\}.$$

Since

$$\lim_{\lambda\to\infty} \frac{b(t)}{s(0)} = (1+o(1))(2\pi t)^{-d/2} \qquad \text{a.s.}$$

and

$$\lim_{\lambda\to\infty} \mathbf{P}\left\{\frac{k(t)-\lambda}{\lambda^{1/2}} < x\right\} \in \mathcal{N}\left(-\frac{\lambda^{1/2}(1+o(1))}{(2\pi t)^{d/2}}, 1-(1+o(1))(2\pi t)^{-d/2}\right)$$

we have

$$\lim_{\lambda\to\infty} Q = \lim_{\lambda\to\infty} \mathbf{P}\left\{\frac{k(t)-\lambda}{\lambda^{1/2}} + \frac{(\lambda^{1/2}+y)(1+o(1))}{(2\pi t)^{d/2}} < x\right\} \in$$
$$\in \mathcal{N}\left(\frac{y(1+o(1))}{(2\pi t)^{d/2}}, 1-(1+o(1))(2\pi t)^{-d/2}\right).$$

As a consequence of Lemmas 1.18 and 1.19 we have

THEOREM 1.16 *There exists a Gaussian process $\{\Gamma(t),\ t \ge 0\}$ with*

$$\mathbf{E}\Gamma(t) = 0, \qquad \mathbf{E}\Gamma^2(t) = 1$$

$$\mathbf{E}\Gamma(t_1)\Gamma(t_2) = \frac{1+o(1)}{(2\pi(t_2-t_1))^{d/2}} \quad (0 \le t_1 < t_2 < \infty)$$

such that

$$\left\{\frac{s(t)-\lambda}{\lambda^{1/2}},\ t \ge 0\right\} \xrightarrow{\mathcal{D}} \{\Gamma(t),\ t \ge 0\}$$

as $\lambda \to \infty$.

Similarly one can prove

THEOREM 1.17 *There exists a Gaussian process* $\{\Gamma^{+,-}(t),\ t \geq 0\}$ *with*

$$\mathbf{E}\Gamma^{+,-}(t) = 0, \qquad \mathbf{E}(\Gamma^{+,-}(t))^2 = 1$$

$$\mathbf{E}\Gamma^{+,-}(t_1)\Gamma^{+,-}(t_2) = \frac{1 + o(1)}{(2\pi(t_2 - t_1))^{d/2}} \qquad (0 \leq t_1 < t_2 < \infty)$$

such that

$$\left\{\frac{s^{+,-}(t)}{\lambda^{1/2}},\ t \geq 0\right\} \xrightarrow{\mathcal{D}} \{\Gamma^{+,-}(t),\ t \geq 0\}$$

as $\lambda \to \infty$.

Further

$$\left\{\left(\frac{t}{\mathbf{E}N(t)}\right)^{1/2} N^{+,-}(t),\ t \geq 0\right\} \xrightarrow{\mathcal{D}} \{W(t),\ t \geq 0\}$$

as $\lambda \to \infty$ *where* $W(t)$ *is the standard Wiener process.*

Chapter 2

Extreme value problems

2.1 Introduction

In this Chapter we study the limit properties of

$$\mathcal{S}(T) = \sup_{0\le t\le T} s(t),$$
$$\mathcal{N}(T) = \sup_{0\le t\le T} (N(t+1) - N(t)),$$
$$\mathcal{D}(T) = \sup_{0\le t\le T} (D(t+1) - D(t)).$$

In order to present our main results we define a function $z = z(u)$ as follows:

Let $z = z(u) = z(u,\lambda,\alpha)$ $(u > 0,\ \lambda > 0,\ \alpha \ge 0)$ be the solution of the equation

$$\left(\frac{z(\log u)^{\alpha}}{\lambda e}\right)^{z} = u.$$

On the properties of the function z we present the following trivial

LEMMA 2.1 *For any $\lambda > 0$ and $\alpha \ge 0$ we have*

$$z(u,\lambda,\alpha) = \frac{1}{1+\alpha}\frac{\log u}{\log_2 u} + \frac{1+o(1)}{(1+\alpha)^2}\frac{\log u}{(\log_2 u)^2}\log_3 u \quad as \quad u \to \infty.$$

Our first goal is to study the limit properties of $\mathcal{S}(T)$. We prove our

THEOREM 2.1 *For any $d \ge 1$ and $\lambda > 0$ there exists a r.v. $T_0 = T_0(d,\lambda,\omega) > 0$ such that*

$$z(T,\lambda,0) - C_1 \le \mathcal{S}(T) \le z(T,Q,0) + C_2 \quad a.s. \tag{2.1}$$

for some $Q > \lambda$ if

$$T \ge T_0, \qquad C_1 > \frac{3}{2} + \frac{12}{d}, \qquad C_2 > 1.$$

Consequently (*see Lemma* 2.1) *for any $\varepsilon > 0$*

$$\frac{\log T}{\log_2 T} + (1-\varepsilon)\frac{\log T}{(\log_2 T)^2}\log_3 T \le \mathcal{S}(T) \le \frac{\log T}{\log_2 T} + (1+\varepsilon)\frac{\log T}{(\log_2 T)^2}\log_3 T \ a.s.$$

if $T \ge T_0$.

Our second goal is to study the limit properties of $\mathcal{N}(T)$. We prove our

THEOREM 2.2 *For any $d \geq 1$, $\lambda > 0$ and $\varepsilon > 0$ there exists a r.v. $T_0 = T_0(d, \lambda, \varepsilon, \omega)$ such that*

(a) *in case $d \geq 3$*

$$z(T, \lambda, 0) - C_3 \leq \mathcal{N}(T) \leq z(T, Q, 0) + C_4 \quad a.s. \tag{2.2}$$

for some $Q > \lambda$ if

$$T \geq T_0, \qquad C_3 > 1, \qquad C_4 > 1;$$

(b) *in case $d = 2$*

$$z(T, \lambda, 1) - C_5 \leq \mathcal{N}(T) \leq z(T, Q, 1) + C_6 \quad a.s. \tag{2.3}$$

for some $Q > \lambda$ if

$$T \geq T_0, \qquad C_5 > 1, \qquad C_6 > 1;$$

(c) *in case $d = 1$ there exists a nonnegative, integer valued r.v. Z such that*

$$\lim_{T \to \infty} \mathcal{N}(T) = Z \qquad a.s. \tag{2.4}$$

On the limit properties of $\mathcal{D}(T)$ we have

THEOREM 2.3 *There exist $0 < c_1 \leq c_2 \leq 1$ such that*

$$c_1 = \liminf_{T \to \infty} \psi(T)\mathcal{D}(T) \leq \limsup_{T \to \infty} \psi(T)\mathcal{D}(T) = c_2 \quad a.s. \tag{2.5}$$

where

$$\psi(T) = \frac{\log_2 T}{\log T}.$$

2.2 Extreme values of Poisson type sequences

In order to illuminate our following results we present the following trivial

LEMMA 2.2 *Let Y be a r.v. with*

$$\mathbf{P}\{Y = k\} = \frac{\lambda^k}{k!} e^{-\lambda} \qquad (k = 0, 1, 2, \ldots, \ \lambda > 0).$$

Then we have

$$\left(\frac{e\lambda}{k}\right)^k \frac{e^{-\lambda}}{\sqrt{2\pi k}} \left(1 - \frac{1}{12k}\right) \leq \mathbf{P}\{Y \geq k\} \leq \left(\frac{e\lambda}{k}\right)^k \quad (k = 0, 1, 2, \ldots).$$

THEOREM 2.4 *Let $Y_1, Y_2, \ldots$ be a sequence of r.v.'s with*

$$\mathbf{P}\{Y_n \geq k\} \leq \left(\frac{e\lambda}{k(\log n)^\alpha}\right)^k k^\beta$$

($k = 1, 2, \ldots,\ n = 1, 2, \ldots,\ \alpha \geq 0,\ \beta \geq 0,\ \lambda > 0$). *Then*

$$Y_n \leq z(n, \lambda, \alpha) + C \qquad a.s. \tag{2.6}$$

and

$$\max(Y_1, Y_2, \ldots, Y_n) \leq z(n, \lambda, \alpha) + C \qquad a.s. \tag{2.7}$$

for all but finitely many n if

$$C > \frac{1+\beta}{1+\alpha}.$$

Proof. Clearly by Lemma 2.1 with $z = z(n, \lambda, \alpha)$ we have

$$\begin{aligned}\mathbf{P}\{Y_n \geq z + C\} &\leq \left(\frac{e\lambda}{(z+C)(\log n)^\alpha}\right)^{z+C} (z+C)^\beta \leq \\ &\leq \left(\frac{e\lambda}{z(\log n)^\alpha}\right)^z \left(\frac{e\lambda}{z(\log n)^\alpha}\right)^C (z+C)^\beta = \\ &= \frac{1}{n}\left(\frac{e\lambda}{z(\log n)^\alpha}\right)^C (z+C)^\beta \leq K\frac{1}{n}\frac{(\log_2 n)^{C-\beta}}{(\log n)^{(1+\alpha)C-\beta}}\end{aligned}$$

if $K = K(\alpha, \beta, C, \lambda)$ is big enough. Hence we have (2.6) by Borel – Cantelli lemma. (2.7) is a trivial consequence of (2.6).

THEOREM 2.5 *Let $Y_1, Y_2, \ldots$ be a sequence of r.v.'s with*

$$\mathbf{P}\{Y_n \geq k\} \geq \left(\frac{e\lambda}{k(\log n)^\alpha}\right)^k k^{-\gamma}$$

($k = 1, 2, \ldots,\ n = 2, 3, \ldots,\ \alpha \geq 0,\ \gamma \geq 0,\ \lambda > 0$) *and*

$$\mathbf{P}\{Y_n \geq k, Y_{n+m} \geq k\} \leq \left(1 + \frac{D}{k^\delta}\right) \mathbf{P}\{Y_n \geq k\}\mathbf{P}\{Y_{n+m} \geq k\}$$

($n = 2, 3, \ldots,\ m = 1, 2, \ldots,\ k = 1, 2, \ldots,\ D > 0,\ \delta > 1$) *if $m \geq k^\rho$ ($\rho > 0$). Then*

$$\max(Y_1, Y_2, \ldots, Y_n) \geq z(n, \lambda, \alpha) - C \qquad a.s.$$

for all but finitely many n if

$$C > \frac{1+\gamma+\rho}{1+\alpha}.$$

Proof. Let

$$I(n,k)=\begin{cases}1 & \text{if} \quad Y_n\geq k,\\ 0 & \text{if} \quad Y_n<k.\end{cases}$$

Then

$$\mathbf{E}\sum_{n=a}^{b-1} I(n,k)=\sum_{n=a}^{b-1}\mathbf{P}\{Y_n\geq k\}\geq (b-a)\left(\frac{e\lambda}{k(\log b)^{\alpha}}\right)^{k} k^{-\gamma} \tag{2.8}$$

and

$$\begin{aligned}
\mathbf{E}\left(\sum_{n=a}^{b-1} I(n,k)\right)^2 &= \sum_{n=a}^{b-1}\mathbf{P}\{Y_n\geq k\}+2\mathbf{E}\sum_{a\leq j<\ell<b} I(j,k)I(\ell,k)= \\
&=\sum_{n=a}^{b-1}\mathbf{P}\{Y_n\geq k\}+2\mathbf{E}\sum_{m=1}^{[k^{\rho}]}\sum_{j=a}^{b-m-1} I(j,k)I(j+m,k)+ \\
&\quad +2\mathbf{E}\sum_{m=[k^{\rho}]+1}^{b-a-1}\sum_{j=a}^{b-m-1} I(j,k)I(j+m,k)\leq \\
&\leq \sum_{n=a}^{b-1}\mathbf{P}\{Y_n\geq k\}+2\mathbf{E}\sum_{m=1}^{[k^{\rho}]}\sum_{j=a}^{b-m-1} I(j,k)I(j+m,k)+ \\
&\quad +2\sum_{m=[k^{\rho}]+1}^{b-a-1}\sum_{j=a}^{b-1}\left(1+\frac{D}{k^{\delta}}\right)\mathbf{P}\{Y_j\geq k\}\mathbf{P}\{Y_{j+m}\geq k\}.
\end{aligned}$$

Hence

$$\begin{aligned}
\operatorname{Var}\sum_{n=a}^{b-1} I(n,k) &\leq \sum_{n=a}^{b-1}\mathbf{P}\{Y_n\geq k\}+2\mathbf{E}\sum_{m=1}^{[k^{\rho}]}\sum_{j=a}^{b-m-1} I(j,k)I(j+m,k)+ \\
&\quad +\frac{2D}{k^{\delta}}\sum_{m=[k^{\rho}]+1}^{b-a-1}\sum_{j=a}^{b-1}\mathbf{P}\{Y_j\geq k\}\mathbf{P}\{Y_{j+m}\geq k\}\leq \\
&\leq \sum_{n=a}^{b-1}\mathbf{P}\{Y_n\geq k\}+2k^{\rho}\sum_{n=a}^{b-1}\mathbf{P}\{Y_n\geq k\}+\frac{D}{k^{\delta}}\left(\sum_{n=a}^{b-1}\mathbf{P}\{Y_n\geq k\}\right)^2.
\end{aligned}$$

Let

$$J_{\nu}=\sum_{n=2^{\nu}}^{2^{\nu+1}-1} I(n,z(2^{\nu+1})-C).$$

Then by (2.8) and Lemma 2.1 we have

$$\begin{aligned}
\mathbf{E}J_{\nu} &\geq 2^{\nu}\left(\frac{e\lambda}{(z(2^{\nu+1})-C)(\log 2^{\nu+1})^{\alpha}}\right)^{z(2^{\nu+1})-C}(z(2^{\nu+1})-C)^{-\gamma}\geq \\
&\geq 2^{\nu}\left(\frac{e\lambda}{z(2^{\nu+1})(\log 2^{\nu+1})^{\alpha}}\right)^{z(2^{\nu+1})}\left(\frac{e\lambda(\log 2)^{-\alpha}}{(z(2^{\nu+1})-C)(\nu+1)^{\alpha}}\right)^{-C}(z(2^{\nu+1})-C)^{-\gamma}= \\
&=\frac{1}{2}(e\lambda(\log 2)^{-\alpha})^{-C}(\nu+1)^{\alpha C}(z(2^{\nu+1})-C)^{C-\gamma}\geq (\log\nu)^{-D_2}\nu^{(1+\alpha)C-\gamma}
\end{aligned}$$

if D_2 is big enough. Then by Chebyshev inequality we get

$$\mathbf{P}\{J_\nu = 0\} \leq \mathbf{P}\left\{|J_\nu - \mathbf{E}J_\nu| \geq \frac{1}{2}\mathbf{E}J_\nu\right\} \leq \frac{4\mathrm{Var}J_\nu}{(\mathbf{E}J_\nu)^2} \leq$$

$$\leq \frac{4}{(\mathbf{E}J_\nu)^2}(2(z(2^{\nu+1}) - C)^\rho + 1) \sum_{n=2^\nu}^{2^{\nu+1}-1} \mathbf{P}\{Y_n \geq z(2^{\nu+1}) - C\} +$$

$$+ \frac{4D}{(\mathbf{E}J_\nu)^2} \frac{1}{(z(2^{\nu+1}) - C)^\delta} \left(\sum_{n=2^\nu}^{2^{\nu+1}-1} \mathbf{P}\{Y_n \geq z(2^{\nu+1}) - C\}\right)^2 =$$

$$= \frac{4}{(\mathbf{E}J_\nu)}(2(z(2^{\nu+1}) - C)^\rho + 1) + \frac{4D}{(z(2^{\nu+1}) - C)^\delta} \leq$$

$$\leq (\log \nu)^{D_3} \left(\frac{\nu^\rho}{\nu^{(1+\alpha)C-\gamma}} + \frac{1}{\nu^\delta}\right)$$

if D_3 is big enough. Hence by Borel – Cantelli lemma

$$J_\nu > 0 \qquad \text{a.s.}$$

for all but finitely many ν. Consequently for any ν big enough there exists a r.v. n_ν such that $2^\nu \leq n_\nu \leq 2^{\nu+1} - 1$ and

$$I(n_\nu, z(2^{\nu+1}) - C) = 1.$$

This implies

$$\max(Y_{2^\nu}, Y_{2^\nu+1}, \ldots, Y_{2^{\nu+1}-1}) \geq z(2^{\nu+1}) - C$$

and if $2^\nu \leq n < 2^{\nu+1}$ then

$$\max(Y_1, Y_2, \ldots, Y_n) \geq \max(Y_1, Y_2, \ldots, Y_{2^{\nu-1}}) \geq z(2^\nu) - C.$$

Since by Lemma 2.1 for any $\varepsilon > 0$ if ν is big enough

$$z(2^\nu) - C \geq z(n) - C - \varepsilon,$$

we have Theorem 2.5.

Observe that

$$z(2^k) \geq z(2^{k+\ell}) - \varepsilon \quad \text{if} \quad \ell \leq \varepsilon \log k.$$

Then Theorem 2.5 easily implies

THEOREM 2.6 *Let $Y_1, Y_2, \ldots$ be a sequence of r.v.'s satisfying the conditions of Theorem 2.5. Then for any $\varepsilon > 0$ and n big enough there exists a sequence of positive, integer valued r.v.'s*

$$\mu_1 = \mu_1(n, \varepsilon, \omega) < \mu_2 = \mu_2(n, \varepsilon, \omega) \ldots < \mu_\xi = \mu_\xi(n, \varepsilon, \omega) \leq n$$

such that

$$Y_{\mu_i} \geq z(n,\lambda,\alpha) - C - \varepsilon \qquad i = 1,2,\ldots,\xi = \xi(n,\varepsilon,\omega),$$
$$\xi \geq \varepsilon \log\log n$$

if

$$C > \frac{1+\gamma+\rho}{1+\alpha}.$$

2.3 Lemmas

LEMMA 2.3 *Let $B_1, B_2, \ldots$ be a sequence of independent r.v.'s with*

$$\mathbf{P}\{B_n = k\} = \binom{f(n)}{k}(g(n))^k(1-g(n))^{n-k} \qquad (0 \leq k \leq f(n);\ n \geq 1)$$

where $f(n) \nearrow \infty$ is an integer valued function, $0 < g(n) < 1$, $g(n) \searrow 0$ and

$$\mu = \sum_{n=1}^{\infty} f(n)g(n) < \infty.$$

Then there exists a $Q = Q(\mu) > 0$ such that

$$\mathbf{P}\left\{\sum_{n=1}^{\infty} B_n \geq k\right\} \leq \left(\frac{Q}{k}\right)^k.$$

Proof. Since as $n \to \infty$

$$\mathbf{E}e^{tB_n} = (1+g(n)(e^t-1))^{f(n)} \sim \exp(g(n)f(n)(e^t-1)),$$

we have

$$\mathbf{E}\exp\left(t\sum_{n=1}^{\infty} B_n\right) = \prod_{n=1}^{\infty}(1+g(n)(e^t-1))^{f(n)} \leq C\exp(\mu(e^t-1))$$

if C is big enough. Hence

$$\mathbf{P}\left\{\sum_{n=1}^{\infty} B_n \geq k\right\} \leq \frac{C\exp(\mu(e^t-1))}{e^{tk}}.$$

Let $t = \log(k/\mu)$. Then

$$\mathbf{P}\left\{\sum_{n-1}^{\infty} B_n \geq k\right\} \leq C\exp\left(\mu\left(\frac{k}{\mu}-1\right) - k\log\frac{k}{\mu}\right) =$$
$$= Ce^{-\mu}\left(\frac{\exp(1+\log\mu)}{k}\right)^k$$

and we have Lemma 2.3.

Let $y_1, y_2, \ldots \in \mathbb{R}^d$ be a deterministic sequence. We say that it is a *D^*-sequence* if

$$T_R = \#\{i:\ R^2 \le \|y_i\| < (R+1)^2\} \le D^*(R+1)^{2d-1} \qquad (R = 0, 1, 2, \ldots).$$

Consider the processes

$$Y_i(t) = y_i + W_i(t) \qquad (i = 1, 2, \ldots;\ t \ge 0)$$

where $W_1(t), W_2(t), \ldots$ are independent Wiener processes and $\{y_i\}$ is a D^*-sequence. Let

$$\begin{aligned}
\tilde{s}(t) &= \#\{i:\ Y_i(t) \in C\},\\
\tilde{S}(T) &= \sup_{0\le t\le T} \tilde{s}(t),\\
U(T) &= \sup_{T\le t<T+1} \tilde{s}(t),\\
V(R,T) &= \#\{i:\ R^2 \le \|y_i\| < (R+1)^2,\ \exists t:\ T \le t < T+1,\ Y_i(t) \in C\}.
\end{aligned}$$

LEMMA 2.4 *There exists a constant $\overline{Q} = \overline{Q}(D^*) > 0$ such that*

$$\mathbf{P}\{U(T) \ge k\} \le \left(\frac{\overline{Q}}{k}\right)^k \qquad (k = 1, 2, \ldots)$$

for any $T \ge 0$.

Proof. Let $\|y_i\| \ge R^2$. Then clearly

$$\mathbf{P}\{\exists t:\ T \le t < T+1,\ Y_i(t) \in C\} \le O\left(T^{-d/2}\exp\left(-\frac{R^4}{2T}\right)\right).$$

Consequently

$$\mathbf{P}\{V(R,T) = k\} = \binom{F}{k} G^k (1-G)^{F-k}$$

where

$$\begin{aligned}
F &= F(R,T) \le D^*(R+1)^{2d-1},\\
G &= G(R,T) \le O\left(T^{-d/2}\exp\left(-\frac{R^4}{2T}\right)\right).
\end{aligned}$$

Since

$$\sum_{R=0}^{\infty} F(R,T)G(R,T) \le A$$

where $A = A(D^*)$ does not depend on T and

$$U(T) \leq \sum_{R=0}^{\infty} V(R,T),$$

by Lemma 2.3 we obtain Lemma 2.4.

For any $0 \leq u < v < \infty$ and for the Borel sets $\mathcal{A} \subset \mathbb{R}^d$, $\mathcal{B} \subset \mathbb{R}^d$ let

$$Y(u,v;\mathcal{A},\mathcal{B}) = \#\{i : i \geq 1, P_i(u) \in \mathcal{A}, P_i(v) \in \mathcal{B}\}$$

where $P_i(\cdot)$ is the process defined in Section 1.2.

Then we have

LEMMA 2.5 *$Y(u,v;\mathcal{A},\mathcal{B})$ is a Poisson r.v. with parameter*

$$\mu(u,v;\mathcal{A},\mathcal{B}) = \frac{\lambda}{(2\pi(v-u))^{d/2}} \int_{\mathcal{A}} \int_{\mathcal{B}} \exp\left(-\frac{\|y-x\|^2}{2(v-u)}\right) dy dx.$$

Proof. Clearly

$$\mathbf{P}\{P_i(v) \in \mathcal{B} \mid P_i(u) = x\} = \frac{1}{(2\pi(v-u))^{d/2}} \int_{\mathcal{B}} \exp\left(-\frac{\|y-x\|^2}{2(v-u)}\right) dy.$$

Let $K(x,dx)$ be the ball around $x \in \mathbb{R}^d$ with volume dx. Then $s(K(x,dx),u)$ is a Poisson r.v. with parameter λdx. Hence we have Lemma 2.5.

Let

$$\mathcal{C}(R) = \mathcal{C}(0,R) = \{x : x \in \mathbb{R}^d, \|x\| \leq R\}.$$

Then a trivial calculation implies

LEMMA 2.6

$$\begin{aligned}
&\lambda C_1 t^{-d/2} \leq \mu(0,t;\mathcal{C},\mathcal{C}) \leq \lambda C_2 t^{-d/2},\\
&\lambda(1 - C_2 t^{-d/2}) \leq \mu(0,t;\overline{\mathcal{C}},\mathcal{C}) \leq \lambda(1 - C_1 t^{-d/2}),\\
&\lambda C_1 e^{-R^2/2} \leq \mu(0,1;\mathcal{C},\overline{\mathcal{C}(R)}) \leq \lambda C_2 e^{-R^2/2},\\
&\lambda C_1 (R t^{-1/2})^d \leq \mu(0,t;\mathcal{C}(R),\mathcal{C}) \leq \lambda C_2 (R t^{-1/2})^d \text{ if } R \leq O(t^{1/2}),\\
&\lambda C_1 t^{-d/2} \exp\left(-\frac{R^2}{2t}\right) \leq \mu(0,t;\overline{\mathcal{C}(R)},\mathcal{C}) \leq \lambda C_2 t^{-d/2} \exp\left(-\frac{R^2}{2t}\right)
\end{aligned}$$

where $0 < C_1 = C_1(d) < C_2 = C_2(d) < \infty$ are positive constants depending only on d and for any Borel set $\mathcal{A} \subset \mathbb{R}^d$ the set $\overline{\mathcal{A}} = \mathbb{R}^d - \mathcal{A}$ is the complementer of $\mathcal{A}$.

Since

$$C_1 t^{-d/2} \leq \mathbf{P}\{P_i(t) \in \mathcal{C} \mid P_i(0) \in \mathcal{C}\} \leq C_2 t^{-d/2}$$

we get

LEMMA 2.7 *For any $k = 0,1,2,\ldots,n$; $n = 1,2,\ldots$ we have*

$$\binom{n}{k}(C_1t^{-d/2})^k(1-C_2t^{-d/2})^{n-k} \le$$
$$\le \mathbf{P}\{Y(0,t;\mathcal{C},\mathcal{C}) = k \mid s(0) = n\} \le \binom{n}{k}(C_2t^{-d/2})^k(1-C_1t^{-d/2})^{n-k}.$$

LEMMA 2.8 *Let*

$$k \le \ell \le 4k, \qquad t^{d/2} \ge k^6.$$

Then

$$\mathbf{P}\{s(0)=k,\ s(t)=\ell\} \le (1+Ck^{-4})\mathbf{P}\{s(0)=k\}\mathbf{P}\{s(t)=\ell\}.$$

Proof. Since

(i) $Y(0,t;\overline{\mathcal{C}},\mathcal{C})$ and $s(0)$ are independent,

(ii) $s(t) = Y(0,t;\mathcal{C},\mathcal{C}) + Y(0,t;\overline{\mathcal{C}},\mathcal{C})$

by Lemmas 2.5 and 2.7 we have

$$\mathbf{P}\{s(0)=k,\ s(t)=\ell\} =$$
$$= \mathbf{P}\{s(0)=k\}\mathbf{P}\{Y(0,t;\mathcal{C},\mathcal{C}) + Y(0,t;\overline{\mathcal{C}},\mathcal{C}) = \ell \mid s(0)=k\} \le$$
$$\le \mathbf{P}\{s(0)=k\}\sum_{j=0}^{k}\binom{k}{j}(C_2t^{-d/2})^j(1-C_1t^{-d/2})^{k-j}\frac{\mu^{\ell-j}}{(\ell-j)!}e^{-\mu} =$$
$$= \mathbf{P}\{s(0)=k\}e^{-\mu}(1-C_1t^{-d/2})^k\mu^\ell\sum_{j=0}^{k}\binom{k}{j}\frac{1}{(\ell-j)!}\left(\frac{C_2t^{-d/2}}{\mu(1-C_1t^{-d/2})}\right)^j.$$

where $\mu = \mu(0,t;\overline{\mathcal{C}},\mathcal{C})$. Our conditions imply

$$(\ell-j)!(t^{d/2})^j \ge (\ell-j)!\ell^{6j}4^{-6j} \ge \ell!\ell^{5j}4^{-6j}$$

hence

$$\mathbf{P}\{s(0)=k,\ s(t)=\ell\} \le$$
$$\le \mathbf{P}\{s(0)=k\}e^{-\mu}\frac{\mu^\ell}{\ell!}(1-C_1t^{-d/2})^k\sum_{j=0}^{k}\binom{k}{j}\left(\frac{C_24^6}{\mu(1-C_1t^{-d/2})\ell^5}\right)^j =$$
$$= \mathbf{P}\{s(0)=k\}e^{-\mu}\frac{\mu^\ell}{\ell!}(1-C_1t^{-d/2})^k\left(1+\frac{C_24^6}{\mu(1-C_1t^{-d/2})\ell^5}\right)^k.$$

Observe that by Lemma 2.6

$$e^{-\mu}\frac{\mu^\ell}{\ell!} \le e^{-\lambda}e^{\lambda C_2t^{-d/2}}\lambda^\ell(1-C_1t^{-d/2})^\ell\frac{1}{\ell!} \le$$
$$\le \mathbf{P}\{s(t)=\ell\}\exp(\lambda C_2t^{-d/2}).$$

Hence

$$\begin{aligned}
&\mathbf{P}\{s(0)=k,\ s(t)=\ell\}\le\\
&\le \mathbf{P}\{s(0)=k\}\mathbf{P}\{s(t)=\ell\}\left(1+\frac{C_2 4^6}{\mu(1-C_1t^{-d/2})\ell^5}\right)^k \exp(\lambda C_2 t^{-d/2})\le\\
&\le \mathbf{P}\{s(0)=k,\ s(t)=\ell\}\left(1+\frac{C}{\ell^5}\right)^k\left(1+\frac{C}{t^{d/2}}\right)
\end{aligned}$$

and we have Lemma 2.8.

LEMMA 2.9 *Let* $t^{d/2}\ge k^6$. *Then*

$$\begin{aligned}
&\mathbf{P}\{s(0)\ge k,\ s(t)\ge k\}\le\\
&\le(1+Ck^{-4})\mathbf{P}\{s(0)\ge k\}\mathbf{P}\{s(t)\ge k\}=(1+Ck^{-4})(\mathbf{P}\{s(0)\ge k\})^2
\end{aligned}$$

Proof.

$$\begin{aligned}
&\mathbf{P}\{s(0)\ge k,\ s(t)\ge k\}=\\
&=\sum_{u=k}^{\infty}\sum_{v=k}^{\infty}\mathbf{P}\{s(0)=u,\ s(t)=v\}=\sum_{u=k}^{4k}\sum_{v=k}^{4k}\mathbf{P}\{s(0)=u,\ s(t)=v\}+\\
&\quad+\sum_{u=4k}^{\infty}\sum_{v=k}^{\infty}\mathbf{P}\{s(0)=u,\ s(t)=v\}+\sum_{u=k}^{4k}\sum_{v=4k}^{\infty}\mathbf{P}\{s(0)=u,\ s(t)=v\}\le\\
&\le\sum_{u=k}^{4k}\sum_{v=k}^{4k}\mathbf{P}\{s(0)=u,\ s(t)=v\}+\mathbf{P}\{s(0)\ge 4k\}+\mathbf{P}\{s(t)\ge 4k\}.
\end{aligned}$$

Taking into account that

$$\mathbf{P}\{s(0)=u,\ s(t)=v\}=\mathbf{P}\{s(0)=v,\ s(t)=u\}$$

and applying Lemma 2.8 we get

$$\begin{aligned}
&\mathbf{P}\{s(0)\ge k,\ s(t)\ge k\}\le\sum_{u=k}^{4k}\sum_{v=k}^{4k}(1+Ck^{-4})\mathbf{P}\{s(0)=u\}\mathbf{P}\{s(t)=v\}+\\
&\quad+2\mathbf{P}\{s(0)\ge 4k\}\le(1+Ck^{-4})\mathbf{P}\{s(0)\ge k\}\mathbf{P}\{s(t)\ge k\}+\left(\frac{C}{k}\right)^{4k}
\end{aligned}$$

which implies Lemma 2.9.

LEMMA 2.10 *There exists a constant* $C>0$ *such that*

$$\mathbf{P}\{s(0)=0,\ s(1)=k\}\ge\left(\frac{\lambda C}{k}\right)^k\qquad(k=1,2,\ldots).$$

Proof. Since

$$\mathbf{P}\{s(1) = k \mid s(0) = 0\} = \mathbf{P}\{Y(0, 1, \overline{\mathcal{C}}, \mathcal{C}) = k\},$$

by Lemmas 2.5 and 2.6 we obtain Lemma 2.10.

LEMMA 2.11 *There exists a constant $C > 0$ such that*

$$\mathbf{P}\{s(0) = 0,\ s(1) = k,\ s(2) = 0\} \geq \left(\frac{\lambda C}{k}\right)^k \qquad (k = 1, 2, \ldots).$$

Proof. Because of symmetry reasons Lemma 2.10 clearly implies

$$\mathbf{P}\{s(0) = k,\ s(1) = 0\} = \mathbf{P}\{s(0) = 0,\ s(1) = k\} \geq \left(\frac{\lambda C}{k}\right)^k.$$

Also because of symmetry we have

$$\begin{aligned}
&\mathbf{P}\{s(0) = 0,\ s(1) = k,\ s(2) = 0\} = \\
&= \mathbf{P}\{s(0) = 0,\ s(2) = 0 \mid s(1) = k\}\mathbf{P}\{s(1) = k\} = \\
&= \mathbf{P}\{s(0) = 0 \mid s(1) = k\}\mathbf{P}\{s(2) = 0 \mid s(1) = k\}\mathbf{P}\{s(1) = k\} = \\
&= \frac{\mathbf{P}\{s(0) = 0,\ s(1) = k\}\mathbf{P}\{s(2) = 0,\ s(1) = k\}}{\mathbf{P}\{s(1) = k\}} \geq \left(\frac{\lambda C}{k}\right)^k.
\end{aligned}$$

Hence we have Lemma 2.11.

A simple consequence of Lemma 2.11 is the following:

LEMMA 2.12

$$\left(\frac{C_1}{k}\right)^k \leq \mathbf{P}\{s(0) = 0,\ s(1) \geq k,\ s(2) = 0\} \leq \left(\frac{C_2}{k}\right)^k.$$

Introduce the following notations:

$$\text{(i)} \quad \overline{Y}(s, t; \mathcal{A}, \mathcal{B}) = \#\{i :\ i \geq 1,\ \exists\ s \leq u, v \leq t,\ P_i(u) \in \mathcal{A},\ P_i(v) \in \mathcal{B}\}$$

i.e. $\overline{Y}(s, t; \mathcal{A}, \mathcal{B})$ is the number of those points which move during the time-interval $[s, t]$ from $\mathcal{A}$ to $\mathcal{B}$,

$$\text{(ii)} \quad \Omega_1 = \Omega_1(R) = \{\overline{Y}(0, 2; \mathcal{L}(R), \overline{\mathcal{L}(2R)}) = \overline{Y}(0, 2; \overline{\mathcal{L}(2R)}, \mathcal{L}(R)) = 0\}$$

i.e. the event Ω_1 occurs if and only if no points move from $\mathcal{L}(R)$ to $\overline{\mathcal{L}(2R)}$ or vice-versa in the time interval $[0, 2]$.

LEMMA 2.13

$$\mathbf{P}\{\Omega_1\} \geq 1 - C\exp\left(-\frac{R^2}{3}\right).$$

Proof. Since
$$\mathbf{P}\{\sup_{0\le s<t\le 2}\|W_i(t)-W_i(s)\|\ge R\}\le Ce^{-R^2/2}$$
we easily get Lemma 2.13.

Let
$$\mathcal{F}_1=\mathcal{F}_1(R)=\mathcal{F}\{\pi(\mathcal{A},t):\ 0\le t\le 2,\ \mathcal{A}\in\mathcal{B}(d),\ \mathcal{A}\subset\mathcal{L}(R)\}$$
be the smallest σ-algebra with respect to which the r.v.'s $\pi(\mathcal{A},t)$ are measurable if $0\le t\le 2$ and $\mathcal{A}$ is any Borel subset of the ball $\mathcal{L}(R)$. ($\mathcal{B}(d)$ is the class of the Borel sets of $\mathbb{R}^d$.)

Similarly let
$$\mathcal{F}_2=\mathcal{F}_2(R)=\mathcal{F}\{\pi(\mathcal{A},t):\ 0\le t\le 2,\ \mathcal{A}\in\mathcal{B}(d),\ \mathcal{A}\subset\overline{\mathcal{L}(2R)}\}.$$

The following lemma is trivial.

LEMMA 2.14 *Let $A\in\mathcal{F}_1$, $B\in\mathcal{F}_2$. Then we have*
$$\mathbf{P}\{AB|\Omega_1\}=\mathbf{P}\{A|\Omega_1\}\mathbf{P}\{B|\Omega_1\}.$$

This lemma means that if we know that no points move from $\mathcal{L}(R)$ to $\overline{\mathcal{L}(2R)}$ and vice-versa then the happenings in $\mathcal{L}(R)$ are independent from the happenings in $\overline{\mathcal{L}(2R)}$.

LEMMA 2.15 *Let*
$$E(t,k)=\{s(t-1)=0,\ s(t)\ge k,\ s(t+1)=0\}\quad t\ge 1$$
and
$$m\ge k,\qquad t\ge\exp(\alpha m\log m),\qquad \alpha>\frac{4}{d}.$$
Then we have
$$\mathbf{P}\{E(1,k)E(t,m)\}\le(1+o(1))\mathbf{P}\{E(1,k)\}\mathbf{P}\{E(t,m)\}.$$

Proof.
$$\begin{aligned}\mathbf{P}\{E(1,k)E(t,m)\}&=\mathbf{P}\{E(1,k)E(t,m)\Omega_1\}+\mathbf{P}\{E(1,k)E(t,m)\overline{\Omega}_1\}\le\\&\le\mathbf{P}\{E(1,k)E(t,m)\Omega_1\}+\mathbf{P}\{\overline{\Omega}_1\}.\end{aligned}$$

Let $t>3$ and define
$$L(t,k)=\{Y(2,t-1;\overline{\mathcal{L}(2R)},C)=Y(2,t+1;\overline{\mathcal{L}(2R)},C)=0,\ Y(2,t;\overline{\mathcal{L}(2R)},C)\ge k\}$$
if $k\ge 0$, and
$$L(t,k)=\{Y(2,t-1;\overline{\mathcal{L}(2R)},C)=Y(2,t+1;\overline{\mathcal{L}(2R)},C)=0\}$$

if $k < 0$, and

$$M(t,k) = \{Y(2,t-1;\mathcal{L}(2R),C) = Y(2,t+1;\mathcal{L}(2R),C) = 0,\ Y(2,t;\mathcal{L}(2R),C) = k\}.$$

Then

$$E(t,m) = \sum_{\ell=0}^{\infty} M(t,\ell)L(t,m-\ell).$$

Hence

$$\begin{aligned}\mathbf{P}\{E(1,k)E(t,m)\Omega_1\} &= \mathbf{P}\left\{E(1,k)\Omega_1 \sum_{\ell=0}^{\infty} M(t,\ell)L(t,m-\ell)\right\} \le \\ &\le \mathbf{P}\{E(1,k)\Omega_1 M(t,0)L(t,m)\} + \mathbf{P}\left\{\sum_{\ell=1}^{\infty} M(t,\ell)\right\}.\end{aligned}$$

Observe that by Lemmas 2.5 and 2.6

$$\begin{aligned}\mathbf{P}\left\{\sum_{\ell=1}^{\infty} M(t,\ell)\right\} &\le \mathbf{P}\{Y(2,t;\mathcal{L}(2R),C) \ge 1\} = \\ &= 1 - \exp(-\mu(0,t-2;\mathcal{L}(2R),C)) \le \\ &\le 1 - \exp(-\lambda C_2(2R(t-2)^{-1/2})^d) \le C(Rt^{-1/2})^d\end{aligned}$$

with some $C = C(\lambda, d) > 0$.

By Lemma 2.14 we have

$$\begin{aligned}&\mathbf{P}\{E(1,k)\Omega_1 M(t,0)L(t,m)\} \le \mathbf{P}\{E(1,k)\Omega_1 L(t,m)\} = \\ &= \mathbf{P}\{\Omega_1\}\mathbf{P}\{E(1,k)L(t,m) \mid \Omega_1\} = \mathbf{P}\{\Omega_1\}\mathbf{P}\{E(1,k) \mid \Omega_1\}\mathbf{P}\{L(t,m) \mid \Omega_1\} = \\ &= \mathbf{P}\{E(1,k)\Omega_1\}\mathbf{P}\{L(t,m) \mid \Omega_1\} \le \mathbf{P}\{E(1,k)\}\mathbf{P}\{L(t,m) \mid \Omega_1\},\end{aligned}$$

and by Lemma 2.13

$$\begin{aligned}&\mathbf{P}\{L(t,m) \mid \Omega_1\} \le \\ &\le \frac{\mathbf{P}\{L(t,m)\}}{\mathbf{P}\{\Omega_1\}} \le \frac{\mathbf{P}\{L(t,m)M(t,0)\}}{\mathbf{P}\{\Omega_1\}} + \frac{\mathbf{P}\{\overline{M(t,0)}\}}{\mathbf{P}\{\Omega_1\}} \le \\ &\le \frac{\mathbf{P}\{E(t,m)\}}{\mathbf{P}\{\Omega_1\}} + \frac{\mathbf{P}\{\overline{M(t,0)}\}}{\mathbf{P}\{\Omega_1\}} \le \mathbf{P}\{E(t,m)\}\left(1 - C\exp\left(-\frac{R^2}{3}\right)\right)^{-1} + \\ &\quad + \frac{1}{\mathbf{P}\{\Omega_1\}} \sum_{i=-1}^{1} \mathbf{P}\{Y(2,t+i;\mathcal{L}(2R),C) \ge 1\} \le \\ &\le \mathbf{P}\{E(t,m)\}\left(1 - C\exp\left(-\frac{R^2}{3}\right)\right)^{-1} + \left(1 - C\exp\left(-\frac{R^2}{3}\right)\right)^{-1} C(Rt^{-1/2})^d.\end{aligned}$$

Hence we have

$$\mathbf{P}\{E(1,k)E(t,m)\} \leq$$
$$\leq C\exp\left(-\frac{R^2}{3}\right) + (CRt^{-1/2})^d + \left(1 - C\exp\left(-\frac{R^2}{3}\right)\right)^{-1} C(Rt^{-1/2})^d +$$
$$+ \mathbf{P}\{E(1,k)\}\mathbf{P}\{E(t,m)\}\left(1 - C\exp\left(-\frac{R^2}{3}\right)\right)^{-1}.$$

Choose $R = m$ (say). Since $m \geq k$ by Lemma 2.12 we get

$$(CRt^{-1/2})^d = o(\mathbf{P}\{E(1,k)\}\mathbf{P}\{E(t,m)\})$$

and

$$\exp\left(-\frac{R^2}{3}\right) = o(\mathbf{P}\{E(1,k)\}\mathbf{P}\{E(t,m)\}).$$

Hence we obtain Lemma 2.15.

2.4 Proofs of Theorems 2.1, 2.2 and 2.3

PROPOSITION 2.1 *Let $y_1, y_2, \ldots \in \mathbb{R}^d$ be a D^*-sequence and consider the processes*

$$\begin{aligned} Y_i(t) &= y_i + W_i(t), \\ \tilde{s}(t) &= \#\{i : Y_i(t) \in C\}, \\ \tilde{S}(T) &= \sup_{0 \leq t \leq T} \tilde{s}(t). \end{aligned}$$

Then for any $d \geq 1$ and $D^* > 0$ there exist a $Q = Q(D^*, d) > 0$ and a r.v. $T_0 = T_0(D^*, d)$ such that

$$\tilde{S}(T) \leq z(T, Q, 0) + C_2 \tag{2.9}$$

for all $T \geq T_0$ and $C_2 > 1$.

Proof. Let

$$k_T = k_T(Q) = z(T, Q, 0) + C_2$$

and observe that for any $\overline{Q} > 0$ and $\varepsilon > 0$

$$\left(\frac{k_T}{\overline{Q}}\right)^{k_T} \geq T(\log T)^{1+\varepsilon}$$

if T and Q are big enough. By Lemma 2.4

$$\mathbf{P}\{U(T) \geq k_T\} \leq \left(\frac{\overline{Q}}{k_T}\right)^{k_T} \leq \frac{1}{T(\log T)^{1+\varepsilon}}$$

and we have (2.9) by the Borel - Cantelli lemma.

Let $\nu = \{Y_1, Y_2, \ldots\}$ be a point process on $\mathbb{R}^d$. We say that it is a *D^*-process* if there exist a $D^* > 0$ and a r.v. $R_0 = R_0(\omega)$ such that for any $R \geq R_0$ we have

$$T_R(\omega) = \#\{i : R^2 \leq \|Y_i\| < (R+1)^2\} \leq D^*(R+1)^{2d-1}. \tag{2.10}$$

Proposition 2.1 implies

PROPOSITION 2.2 *Let $Y_1, Y_2, \ldots \in \mathbb{R}^d$ be a D^*-process and consider the processes*

$$\begin{aligned} \overline{Y}_i(t) &= Y_i + W_i(t), \\ \overline{s}(t) &= \#\{i : \overline{Y}_i(t) \in C\}, \\ \overline{S}(T) &= \sup_{0 \leq t \leq T} \overline{s}(t). \end{aligned}$$

Then for any $d \geq 1$ and $D^* > 0$ there exist a $Q = Q(D^*, d) > 0$ and a r.v. $T_0 = T_0(D^*, d)$ such that

$$\overline{S}(T) \leq z(T, Q, 0) + C_2 \tag{2.11}$$

for all $T \geq T_0$.

Since a Poisson process is a D^*-process we have the upper part of (2.1) by (2.11). Now we turn to the proof of the lower part of (2.1).
Apply Theorem 2.5 with

$$Y_i = s(i), \qquad \alpha = 0, \qquad \delta = 4, \qquad \rho = 12/d, \qquad \gamma = 1/2$$

by Lemmas 2.9 and 2.2 we obtain the lower part of (2.1). Hence Theorem 2.1 is proved.
Applying Theorem 2.6 just like Theorem 2.5 above we obtain

PROPOSITION 2.3 *For any $d \geq 1$, $\lambda > 0$ and T big enough there exists a sequence of positive, integer valued r.v.'s*

$$\mu_1 < \mu_2 < \ldots < \mu_\xi \leq T$$

such that

$$\begin{aligned} s(\mu_i) &\geq z(T, \lambda, 0) - C \qquad i = 1, 2, \ldots, \xi, \\ \xi &\geq \varepsilon \log\log T \end{aligned}$$

if

$$C > \frac{3}{2} + \frac{12}{d}$$

and $\varepsilon = \varepsilon(C) > 0$ is small enough.

LEMMA 2.16 *For any $\lambda > 0$ there exists a $Q = Q(\lambda, d) > 0$ such that*

$$\mathbf{P}\{N(1) \geq k\} \leq \left(\frac{eQ}{k}\right)^k \qquad (k = 1, 2, \ldots).$$

Proof. It is a trivial consequence of Lemma 2.2 and Theorem 1.3.

LEMMA 2.17 *For any $\lambda > 0$ there exists a $Q = Q(\lambda, d) > 0$ such that*

$$\left(\frac{e\lambda}{k}\right)^k \frac{e^{-\lambda}}{\sqrt{2\pi k}}\left(1 - \frac{1}{12k}\right) \leq \mathbf{P}\{\mathcal{N}(1) \geq k\} \leq \left(\frac{eQ}{k}\right)^k.$$

Proof. Since $\mathcal{N}(1) \leq N(2)$ and $\mathcal{N}(1) \geq N(0) = s(0)$ by Lemma 2.2 and Theorem 1.3 we get Lemma 2.17.

LEMMA 2.18 *For any $\lambda > 0$, $d \geq 3$ there exist a $Q = Q(\lambda, d) > 0$ and a $C_d > 0$ such that*

$$\left(\frac{e\lambda C_d}{k}\right)^k \frac{e^{-\lambda C_d}}{\sqrt{2\pi k}} \leq \mathbf{P}\{N(t+1) - N(t) \geq k\} \leq \left(\frac{eQ}{k}\right)^k \qquad (t = 0, 1, 2, \ldots).$$

Proof. Observe that by Theorem 1.3 for any $d \geq 3$

$$\mathbf{E}(N(t+1) - N(t)) = \lambda C_d + o(1).$$

Hence by Lemma 2.2 we have Lemma 2.18.

Apply Theorem 2.4 with

$$Y_i = \sup_{i \leq t \leq i+1} (N(t+1) - N(t)), \qquad (i = 1, 2, \ldots, [T])$$

$$\alpha = 0, \qquad \beta = 0, \qquad \lambda = Q$$

by Lemma 2.18 we obtain the upper part of (2.2). Applying Theorem 2.5 with

$$Y_i = \sup_{2i \leq t \leq 2i+1} (N(t+1) - N(t)) \qquad (i = 1, 2, \ldots, [T/2])$$

$$\alpha = 0, \qquad \gamma = 1/2, \qquad n = [T/2]$$

by Lemma 2.18 we obtain the lower part of (2.2).

Repeating the above proof with $\alpha = 1$ and $d = 2$ we obtain (2.3).

By Lemma 2.2 and Theorem 1.3 for $d = 1$ we have

$$N(t+1) - N(t) \leq 3 \qquad \text{a.s.}$$

if t is big enough. Since $\mathcal{N}(t)$ is nondecreasing we have (2.4). Hence Theorem 2.2 is proved.

Since $\mathcal{D}(T) \leq \mathcal{S}(T)$ by the zero-one law and Theorem 2.1 we have the upper part of (2.5).

Clearly for any $0 < \varepsilon < 1/2$ there exist a $q = q(\varepsilon) > 0$ and a $p = p(\varepsilon) > 0$ such that

$$\mathbf{P}\left\{\inf_{t \leq u \leq t+1} s(u) \geq q\frac{\log T}{\log_2 T} \,\middle|\, s(t) \geq (1-\varepsilon)\frac{\log T}{\log_2 T}\right\} \geq p.$$

Hence we have the lower part of (2.5) by the zero-one law and Theorem 2.3 is proved.

2.5 Exploison in $\mathcal{C}$

Theorem 2.1 told us that for some $T > 0$ the number of particles in $\mathcal{C}$ (i.e. $s(T)$) can be very big. We are also interested in the question: how rapidly (or slowly) can $s(\cdot)$ achieve its large values. Our following Theorem claims that it sometimes happens like an exploison.

THEOREM 2.7 *For any*

$$0 < L < \frac{d}{2(2+d)}$$

there exists a random sequence of integers $0 < n_1 = n_1(\omega, L) < n_2 = n_2(\omega, L) < \ldots$ *such that*

$$s(n_i - 1) = 0, \qquad s(n_i) \geq L\frac{\log n_i}{\log\log n_i}, \qquad s(n_i + 1) = 0.$$

Proof. Let

$$k = k(n) = L\frac{\log n}{\log\log n}.$$

Then for any $C > 0$

$$\left(\frac{C}{k}\right)^k = \exp(-k(\log k - \log C)) \geq \frac{1}{2n^L}$$

if n is big enough. Hence by Lemma 2.11

$$\sum_{n=1}^{N} \mathbf{P}\{E(n,\ k(n))\} \geq CN^{1-L}$$

if N is big enough and $C > 0$ is small enough. ($E(\cdot,\cdot)$ is defined before Lemma 2.15.)

Let $\alpha L < \beta < 1 - 2L$ where $\alpha > 4d^{-1}$ is the constant of Lemma 2.15. Note that the inequality $\alpha L < \beta < 1 - 2L$ can hold only if $(\alpha + 2)L < 1$. However, for any

$$L < \frac{d}{2(2+d)}$$

α can be chosen such that

$$\frac{4}{d} < \alpha < \frac{1}{L}.$$

Consider

$$\sum_{1 \le i < j \le N} \mathbf{P}\{E(i,\ k(i))E(j,\ k(j))\} =$$

$$= \sum_{i=1}^{N} \sum_{j=i+1}^{i+i^\beta} \mathbf{P}\{E(i,\ k(i))E(j,\ k(j)\} + \sum_{i=1}^{N} \sum_{j=i+i^\beta+1}^{N} \mathbf{P}\{E(i,\ k(i))E(j,\ k(j))\} \le$$

$$\le \sum_{i=1}^{N} i^\beta + \sum_{i=1}^{N} \sum_{j=i+i^\beta+1}^{N} \mathbf{P}\{E(i,\ k(i))E(j,\ k(j))\} \le$$

$$\le O(N^{\beta+1}) + \sum_{i=1}^{N} \sum_{j=i+i^\beta+1}^{N} \mathbf{P}\{E(i,\ k(i))E(j,\ k(j))\}.$$

Then by Lemma 2.15

$$\sum_{i=1}^{N} \sum_{j=i+i^\beta+1}^{N} \mathbf{P}\{E(i,\ k(i))E(j,\ k(j))\} \le \sum_{1 \le i < j \le N} (1+o(1))\mathbf{P}\{E(i,\ k(i))\}\mathbf{P}\{E(j,\ k(j))\}$$

and we obtain Theorem 2.7 by Borel - Cantelli lemma (cf. [44] Borel - Cantelli lemma 2* p. 27).

2.6 Charged particles

Taking into account again the independence of the processes $s^+(t)$ and $s^-(t)$ (cf. Section 1.6) we get without any difficulty the following:

THEOREM 2.8 *For any $d \ge 1$, $\lambda > 0$ there exists a r.v. $T_0 = T_0(d, \lambda, \omega) > 0$ such that*

$$z(T, \lambda/2, 0) - C_1 \le S^{+,-}(T) \le z(T, Q, 0) + C_2 \qquad a.s.$$

for some $Q > \lambda/2$ if

$$T \ge T_0, \qquad C_1 > \frac{3}{2} + \frac{12}{d}, \qquad C_2 > 1.$$

Theorems 2.2 and 2.3 can be also extended to the processes

$$\mathcal{N}^{+,-}(T) = \sup_{0 \le t \le T} N^{+,-}(t)$$

and

$$\mathcal{D}^{+,-}(T) = \sup_{0 \le t \le T} D^{+,-}(t).$$

Chapter 3

Changing the initial process and the motion

3.1 Introduction

In this Chapter we intend to show that the main results of Chapters 1 and 2 remain true by slightly changing the initial process and /or the motion. We start with the initial process. We define the "λ-homogeneous" process and prove that the above results remain true replacing the initial Poisson process by a λ-homogeneous process.

Let ν be a point process on $\mathbb{R}^d$. Denote its points by $Y_1, Y_2, \dots$ in an arbitrary order. We say that ν is *λ-homogeneous* if there exists a one-to-one mapping $f(\cdot)$ of the elements of the sequence $Y_1, Y_2, \dots$ to $\mathbb{Z}^d$ such that

$$\mathbf{P}\{\|\lambda^{-1/d} f(Y_i) - Y_i\| \geq \|f(Y_i)\|^{1/2+\varepsilon} \mid f(Y_i) = x \in \mathbb{Z}^d\} \leq$$

$$\leq C_1 \exp(-C_2 \|x\|^{\varepsilon}) \tag{3.1}$$

for some $0 < \varepsilon < 1/2$, $C_1 > 0$, $C_2 > 0$.

In order to explain the meaning of condition (3.1) note that if $\dots < Y_{-1} < 0 < Y_1 < \dots$ is a Poisson process on $\mathbb{R}^1$ of intensity λ then

$$\mathbf{P}\left\{\left|Y_i - \frac{i}{\lambda}\right| \geq |i|^{1/2+\varepsilon}\right\} \leq C_1 \exp(-C_2 |i|^{\varepsilon}) \tag{3.2}$$

i.e. (3.1) is satisfied with $f(Y_i) = i$.

We also note that it is easy to see that (3.1) is satisfied in the following cases:

(i) ν is a Poisson process on $\mathbb{R}^d$ of intensity λ,

(ii) ν is the set of all points of $\mathbb{Z}^d$ and $\lambda = 1$,

(iii) if ε_x $(x \in \mathbb{Z}^d)$ is an array of i.i.d.r.v's with

$$\mathbf{P}\{\varepsilon_x = 1\} = 1 - \mathbf{P}\{\varepsilon_x = 0\} = p \qquad (0 < p < 1)$$

and x is an element of ν if $\varepsilon_x = 1$ $\quad (\lambda = p)$,

(iv) if $\varepsilon(x)$ $(x \in \mathbb{Z}^d)$ is an array of i.i.d.r.v's with

$$\mathbf{P}\{\varepsilon(x) = k\} = \frac{\mu^k}{k!} e^{-\mu} \qquad (\mu > 0;\ k = 0, 1, 2, \ldots)$$

and the elements of ν are $Y_1(x) = Y_2(x) = \ldots = Y_{\varepsilon(x)} = x$ $(x \in \mathbb{Z}^d)$ and $\lambda = \mu$.

We saw that a Poisson process on $\mathbb{R}^1$ of intensity λ is λ-homogeneous. In order to see that it is so on $\mathbb{R}^d$ too, let $Y_i = (Y_{i1}, Y_{i2}, \ldots, Y_{id})$ and consider the cylinder set

$$C(x_1, x_2, \ldots, x_{d-1}) = [x_1, x_1 + 1) \times \cdots \times [x_{d-1}, x_{d-1} + 1) \times \mathbb{R}^1$$

where $x_1, x_2, \ldots, x_{d-1}$ are integers. Consider those elements of the sequence $\{Y_i\}$ which are located in

$$\begin{aligned} &\lambda^{-1/d} C(x_1, \ldots, x_{d-1}) = \\ &= [\lambda^{-1/d} x_1, \lambda^{-1/d}(x_1 + 1)) \times \cdots \times [\lambda^{-1/d} x_{d-1}, \lambda^{-1/d}(x_{d-1} + 1)) \times \mathbb{R}^1 \end{aligned}$$

and order them by the magnitude of Y_{id} i.e.

$$\ldots < Y_{-1d} < 0 < Y_{1d} < \ldots \tag{3.3}$$

Define the one-to-one mapping $f(\cdot)$ by

$$f(Y_i) = (x_1, x_2, \ldots, x_{d-1}, m(i))$$

if $Y_i \in \lambda^{-1/d} C(x_1, x_2, \ldots, x_{d-1}) \times \mathbb{R}^1$ and $m(i)$ is the index of Y_i in the reordering of (3.3). It is easy to see that this $f(\cdot)$ satisfies (3.1). One can see similarly that in cases (ii), (iii), (iv) ν is also a λ-homogeneous process.

Remark 3.1. Clearly a λ-homogeneous process is a D^*-process.

3.2 Coupling

The following Lemma is trivial.

LEMMA 3.1 *Let* $d = 1$, $x \in \mathbb{R}^1$, $y \in \mathbb{R}^1$, $|x - y| \le t^{1/4+\rho}$, $\rho > 0$,

$$X(t) = x + W_1(t)$$

and

$$Y(t) = y + W_2(t)$$

where $W_1(t)$ *and* $W_2(t)$ *are independent Wiener processes. Then the probability that the processes* $X(\cdot)$ *and* $Y(\cdot)$ *do not meet each other up to* $t^{1-\gamma}$ *is small. In fact*

$$\mathbf{P}\left\{\inf_{0 \le s \le t^{1-\gamma}} |X(s) - Y(s)| \ne 0\right\} \le \mathbf{P}\left\{\sup_{0 \le s \le 1} \sqrt{2} W(s) \le \frac{t^{1/4+\rho}}{t^{1/2-\gamma/2}}\right\} \le O\left(\frac{t^{\rho+\gamma/2}}{t^{1/4}}\right)$$

for any $0 < \gamma < 1$.

Let $\nu = \{Y_1, Y_2, \ldots\}$ be a λ-homogeneous process and $\pi^\lambda = \{X_1, X_2, \ldots\}$ be a Poisson process in $\mathbb{R}^d$

Note that (3.1) easily implies that there exist reorderings of the sequences $\{X_i\}$ and $\{Y_i\}$ such that

$$\mathbf{P}\left\{\bigcup_{\{i:\|X_i\|\leq t^{1/2+\beta}\}} \{\|X_i - Y_i\| \geq t^{1/4+\beta/2+\epsilon}\}\right\} \leq C_1 \exp(-C_2 t^\epsilon) \tag{3.4}$$

for any $d \geq 1$ and $\beta > 0$. In fact ν and π^λ might be reordered such a way that X_i and Y_i are mapped to the same point of $\mathbb{Z}^d$.

Let

$$\begin{aligned}
X_i(t) &= X_i + W_i^\pi(t), \qquad (i = 1, 2, \ldots;\ t \geq 0)\\
Y_i(t) &= Y_i + W_i^\nu(t), \qquad (i = 1, 2, \ldots;\ t \geq 0)\\
\mathcal{N}(t) &= \mathcal{N}(t, K) = \mathcal{N}(t, K, \beta, \gamma, \rho) =\\
&= \#\{i : i \leq t^{(1/2+\beta)K}, \inf_{0\leq s\leq t^{1-\gamma}} |X_i(s) - Y_i(s)| \neq 0, |X_i - Y_i| \leq t^{1/4+\rho}\}
\end{aligned}$$

where $K \geq 1$, $0 < \beta < 1/2$, $0 < \rho < 1/4$, $0 < \gamma < 1$ and $W_1^\pi, W_2^\pi, \ldots; W_1^\nu, W_2^\nu, \ldots$ are independent Wiener processes in $\mathbb{R}^d$.

Now we prove our

LEMMA 3.2 *Let $d = 1$. Then for any $K > 0$ and $\delta > 0$ we have*

$$\mathbf{P}\{\mathcal{N}(t) \geq t^{\Delta+\delta}\} \leq \mathbf{P}\{\mathcal{N}(t) \geq (\mathbf{E}\mathcal{N}(t))O(t^\delta)\} \leq C_1 \exp(-C_2 t^\delta)$$

where

$$\Delta = \left(\frac{1}{2} + \beta\right)K + \rho + \frac{\gamma}{2} - \frac{1}{4}.$$

Proof. By Lemma 3.1

$$\mathbf{E}\mathcal{N}(t) \leq O(t^\Delta)$$

and

$$\mathbf{P}\{\mathcal{N}(t) \geq N\} \leq \sum_{m=N}^{t^{(1/2+\beta)K}} \binom{t^{(1/2+\beta)K}}{m} \left(O\left(\frac{t^{\rho+\gamma/2}}{t^{1/4}}\right)\right)^m \left(1 - O\left(\frac{t^{\rho+\gamma/2}}{t^{1/4}}\right)\right)^{t^{(1/2+\beta)K-m}}$$

Hence

$$\begin{aligned}
\mathbf{E}\exp(\mathcal{N}(t)) &= \left(1 + O\left(\frac{t^{\rho+\gamma/2}}{t^{1/4}}\right)\right)^{t^{(1/2+\beta)K}} \leq\\
&\leq \exp\left(t^{(1/2+\beta)K} O\left(\frac{t^{\rho+\gamma/2}}{t^{1/4}}\right)\right) = \exp(O(1)t^{(1/2+\beta)K+\rho+\gamma/2-1/4})
\end{aligned}$$

which easily implies Lemma 3.2 applying the Markov inequality for the moment generating fuction of $\mathcal{N}(t)$.

For any $d \geq 1$ introduce the following notations:

$$\begin{aligned}
\pi(\mathbb{R}^d, t) &= \{X_1(t), X_2(t), \ldots\},\\
\nu(\mathbb{R}^d, t) &= \{Y_1(t), Y_2(t), \ldots\},\\
X_i(t) &= (X_{i1}(t), X_{i2}(t), \ldots, X_{id}(t)),\\
Y_i(t) &= (Y_{i1}(t), Y_{i2}(t), \ldots, Y_{id}(t)),\\
\tau_{ij} &= \min\{t :\ t \geq 0,\ X_{ij}(t) = Y_{ij}(t)\},\ (i = 1, 2, \ldots;\ j = 1, 2, \ldots, d)\\
\tilde{X}_{ij}(t) &= \begin{cases} X_{ij}(t) & \text{if} \quad t \leq \tau_{ij},\\ Y_{ij}(t) & \text{if} \quad t > \tau_{ij}, \end{cases}\\
\tilde{X}_i(t) &= (\tilde{X}_{i1}(t), \tilde{X}_{i2}(t), \ldots, \tilde{X}_{id}(t)),\\
\mathcal{M}(t) &= \#\{i :\ i \leq t^{(1/2+\beta)d},\ \tilde{X}_i(t^{1-\gamma}) \neq Y_i(t^{1-\gamma})\},\\
\mathcal{L}(t) &= \#\{i :\ i \leq t^{(1/2+\beta)d},\ \|X_i(0) - Y_i(0)\| \geq t^{1/4+\rho}\}.
\end{aligned}$$

Choosing $\rho = \beta/2 + \varepsilon$, by (3.4) we have

$$\mathbf{P}\{\mathcal{L}(t) \geq 1\} \leq C_1 \exp(-C_2 t^{\varepsilon}). \tag{3.5}$$

By (3.5) and Lemma 3.2 (with $K = d,\ \rho = \beta/2 + \varepsilon$) we have

LEMMA 3.3

$$\mathbf{P}\{\mathcal{M}(t) \geq t^{\psi}\} \leq C_1(\exp(-C_2 t^{\delta}) + \exp(-C_2 t^{\varepsilon}))$$

where

$$\psi = \left(\frac{1}{2} + \beta\right) d + \frac{\beta}{2} + \varepsilon + \frac{\gamma}{2} - \frac{1}{4} + \delta.$$

In order to see the meaning of Lemma 3.3 consider the particles located in

$$X_1, X_2, \ldots, X_{t^{(1/2+\beta)d}}$$

at time $t = 0$ and the corresponding particles

$$Y_1, Y_2, \ldots, Y_{t^{(1/2+\beta)d}}.$$

It essentially means that we practically consider the particles located in $C(0, t^{1/2+\beta})$ at time $t = 0$. Then $\mathcal{M}(t)$ is the number of those particles which are located in $C(0, t^{1/2+\beta})$ at time $t = 0$ and which do not meet with the corresponding particles of $\nu(\mathbb{R}^d, t)$ up to $t^{1-\gamma}$. Lemma 3.3 tells us that $\mathcal{M}(t)$ is not more than t^{ψ} where

$$\psi > \left(\frac{1}{2} + \beta\right) d + \frac{\beta}{2} + \frac{\gamma}{2} - \frac{1}{4}.$$

Choose any t^ψ points among the points $\tilde{X}_i(t^{1-\delta})$ $(1 \le i \le t^{(1/2+\beta)d})$. Let $j(1), j(2), \ldots, j(t^\psi)$ be the choosen indices. Then for any n $(2t^{1-\delta} \le n \le t)$

$$\mathbf{P}\{\text{at least } K \text{ among } \tilde{X}_{j(1)}(n), \tilde{X}_{j(2)}(n), \ldots, \tilde{X}_{j(t^\psi)}(n) \text{ are in } \mathcal{C}\} =$$

$$= \sum_{j=K}^{t^\psi} \binom{t^\psi}{j} (O(t^{-(1-\delta)d/2}))^j (1 - O(t^{-(1-\delta)d/2}))^{t^\psi - j} \le O\left(\left(\frac{t^\psi}{t^{(1-\delta)d/2}}\right)^K\right).$$

Hence

$$\mathbf{P}\{\exists n: \ 2t^{1-\delta} \le n \le t \text{ such that at least } K \text{ among the } t^\psi \text{ points are in } \mathcal{C}\} \le$$

$$\le t\, O\left(\left(\frac{t^\psi}{t^{(1-\delta)d/2}}\right)^K\right).$$

Choosing the parameters $\beta, \gamma, \delta, \varepsilon$ small enough $\psi - (1-\delta)d/2$ is close enough to $-1/4$. Hence we have:

among the points $\tilde{X}_{j(1)}(s), \tilde{X}_{j(2)}(s), \ldots, \tilde{X}_{j(t^\psi)}(s)$ not more than 8 can be in $\mathcal{C}$ a.s. if $s \in [2t^{1-\delta}, t]$ and t is big enough. (3.6)

(3.6) tells us that among the particles considered in the definition of $\mathcal{M}(t)$ not more than 8 visit the $\mathcal{C}$ in the time interval $[2t^{1-\delta}, t]$.

3.3 How to apply the coupling?

In this Section we intend to show how can we generalize the results of Chapters 1 and 2 for a λ-homogeneous process by the above coupling results.

Consider the processes

$$X(t) = \{X_1(t), X_2(t), \ldots\},$$
$$Y(t) = \{Y_1(t), Y_2(t), \ldots\}$$

where $X_i(t)$ and $Y_i(t)$ are defined in Section 3.2 and let $N_X(t)$ resp. $N_Y(t)$ be the number of distinct points of $X(\cdot)$ resp. $Y(\cdot)$ which visit $\mathcal{C}$ up to time t, i.e.

$$N_X(t) = \#\{i: \ \exists 0 \le s \le t \text{ such that } X_i(s) \in \mathcal{C}\},$$
$$N_Y(t) = \#\{i: \ \exists 0 \le s \le t \text{ such that } Y_i(s) \in \mathcal{C}\},$$
$$N_{\tilde{X}}(t) = \#\{i: \ \exists 0 \le s \le t \text{ such that } \tilde{X}_i(s) \in \mathcal{C}\}.$$

Similarly let

$$s_X(t) = \#\{i: \ X_i(t) \in \mathcal{C}\},$$
$$s_Y(t) = \#\{i: \ Y_i(t) \in \mathcal{C}\},$$

$$\begin{aligned}
D_X(T) &= \int_0^T s_X(t)dt, \\
D_Y(T) &= \int_0^T s_Y(t)dt, \\
\mathcal{S}_X(T) &= \sup_{0\le t\le T} s_X(t), \\
\mathcal{S}_Y(T) &= \sup_{0\le t\le T} s_Y(t), \\
\mathcal{N}_X(T) &= \sup_{0\le t\le T} (N_X(t+1) - N_X(t)), \\
\mathcal{N}_Y(T) &= \sup_{0\le t\le T} (N_Y(t+1) - N_Y(t)), \\
\mathcal{D}_X(T) &= \sup_{0\le t\le T} (D_X(t+1) - D_X(t)), \\
\mathcal{D}_Y(T) &= \sup_{0\le t\le T} (D_Y(t+1) - D_Y(t)).
\end{aligned}$$

Further let

$$\begin{aligned}
A_X(t) &= \#\{i:\ i \ge t^{(1/2+\beta)d},\ \exists 0\le s\le t \text{ such that } X_i(s)\in \mathcal{C}\}, \\
A_Y(t) &= \#\{i:\ i \ge t^{(1/2+\beta)d},\ \exists 0\le s\le t \text{ such that } Y_i(s)\in \mathcal{C}\}.
\end{aligned}$$

Clearly for any $\beta > 0$

$$A_X(t) = A_Y(t) = 0 \qquad \text{a.s.} \tag{3.7}$$

for all t big enough.

Now we present our

THEOREM 3.1

$$\lim_{t\to\infty} \mathbf{P}\left\{ \frac{N_Y(t) - \mathbf{E}N_Y(t)}{(\mathbf{E}N_Y(t))^{1/2}} < x \right\} = \Phi(x).$$

Proof. Let

$$\begin{aligned}
N_X^\beta(t) &= \#\{i:\ i < t^{(1/2+\beta)d},\ \exists 0\le s\le t:\ X_i(s)\in\mathcal{C}\}, \\
N_Y^\beta(t) &= \#\{i:\ i < t^{(1/2+\beta)d},\ \exists 0\le s\le t:\ Y_i(s)\in\mathcal{C}\}, \\
N_{\tilde X}^\beta(t) &= \#\{i:\ i < t^{(1/2+\beta)d},\ \exists 0\le s\le t:\ \tilde X_i(s)\in\mathcal{C}\}.
\end{aligned}$$

Then by (3.7)

$$N_X(t) = N_X^\beta(t) \quad \text{and} \quad N_Y(t) = N_Y^\beta(t) \quad \text{a.s.}$$

for all t big enough.

Consider

$$\frac{N_X^\beta(t) - \mathbf{E}N_X^\beta(t)}{(\mathbf{E}N_X^\beta(t))^{1/2}} = \frac{N_X^\beta(t) - N_X^\beta(t^{1-\gamma}) - \mathbf{E}(N_X^\beta(t) - N_X^\beta(t^{1-\gamma}))}{(\mathbf{E}N_X^\beta(t))^{1/2}} + \\ + \frac{N_X^\beta(t^{1-\gamma}) - \mathbf{E}N_X^\beta(t^{1-\gamma})}{(\mathbf{E}N_X^\beta(t))^{1/2}}.$$

Since

$$\lim_{t\to\infty} \frac{\mathbf{E}N_X^\beta(t^{1-\gamma})}{\mathbf{E}N_X^\beta(t)} = 0,$$

we have

$$\frac{N_X^\beta(t^{1-\gamma}) - \mathbf{E}N_X^\beta(t^{1-\gamma})}{(\mathbf{E}N_X^\beta(t))^{1/2}} \to 0$$

in probability as $t \to \infty$. Similar result can be obtained for $N_Y^\beta(\cdot)$.

By (3.6) we have

$$|(N_X^\beta(t) - N_X^\beta(t^{1-\gamma})) - (N_Y^\beta(t) - N_Y^\beta(t^{1-\gamma}))| \le 8$$

a.s. for all t big enough. Hence we have Theorem 3.1.

Similarly one can prove

THEOREM 3.2

$$\lim_{t\to\infty} \mathbf{P}\left\{\frac{D_Y(t) - a(t)}{b(t)} < x\right\} = \Phi(x),$$

$$\frac{\log T}{\log_2 T} \le S_Y(T) \le \frac{\log T}{\log_2 T} + (1+\varepsilon)\frac{\log T}{(\log_2 T)^2}\log_3 T$$

a.s. for all T big enough.

3.4 Changing the motion

Going through once more on the proofs of the Theorems of Sections 1.3 and 1.4 we can realize that the condition saying that the motions are governed by Wiener processes is only slightly used. Having an other motion the main difficulty is to find the correct analogue of (1.2). Here we study only the case of simple, symmetric random walks. In fact we consider the following

Model. Let $\{\pi_x,\ x \in \mathbb{Z}^d\}$ be an array of i.i.d. Poisson r.v.'s of parameter λ. Assume that in $x \in \mathbb{Z}^d$ π_x particles are located at time $t = 0$. Hence π_x defines a random field on $\mathbb{Z}^d$ Denote the points of this field by $\{P_1, P_2, \ldots\}$ in any order. Let

$S_1(n), S_2(n), \ldots$ $(n = 0, 1, 2, \ldots)$ be a sequence of independent, simple, symmetric random walks on $\mathbb{Z}^d$ and define

$$P_i(n) = P_i + S_i(n) \qquad (i = 1, 2, \ldots;\ n = 0, 1, 2, \ldots)$$

and

$$P(n) = \{P_1(n), P_2(n), \ldots\}.$$

Clearly $P(0) = \{P_1, P_2, \ldots\}$. Introduce the following notations

$$\begin{aligned} s(x,n) &= \#\{i:\ P_i(n) = x\}, \\ s(0,n) &= s(n), \\ S(n) &= \max_{0 \le k \le n} s(k), \\ N(x,n) &= \#\{i:\ \exists 0 \le k \le n,\ P_i(k) = x\}, \\ N(0,n) &= N(n), \\ D(x,n) &= \sum_{k=0}^{n} s(x,n), \\ D(n) &= D(0,n). \end{aligned}$$

Having this model it is easy to see that the results of Sections 1.3 and 1.4 remain true with slight modifications.

THEOREM 3.3

$$\mathbf{P}\{N(n) = k\} = \frac{\mu^k}{k!} e^{-\mu} \qquad (k = 0, 1, 2, \ldots) \tag{3.8}$$

where

$$\mu = \mu(n) = \mathbf{E}N(n) = \begin{cases} 2\lambda \left(\dfrac{2n}{\pi}\right)^{1/2} (1 + o(1)) & \text{if } d = 1, \\ \dfrac{\lambda \pi n}{\log n} + O\left(\dfrac{n \log\log n}{(\log n)^2}\right) & \text{if } d = 2, \\ K_3 n + O(n^{1/2}) & \text{if } d = 3, \\ K_4 n + O(\log n) & \text{if } d = 4, \\ K_d n + \beta_d + O(n^{2-d/2}) & \text{if } d \ge 5 \end{cases} \tag{3.9}$$

where β_d's $(d = 5, 6, \ldots)$ are positive constants and

$$K_d = \mathbf{P}\{S_k \ne 0,\ k = 1, 2, \ldots\} \qquad (d = 3, 4, \ldots).$$

Proof. Let $x \in \mathbb{Z}^d$ and

$$p(x,n) = \mathbf{P}\{\exists 0 \le k \le n \text{ for which } P_i(k) = 0 \mid P_i(0) = x\}$$

be the probability that a point starting from x visits $0 \in \mathbb{Z}^d$ during the first n steps. Note that $p(0, n) = 1$.

Then repeating step by step the proof of Theorem 1.3 we get (3.8) with

$$\mu = \lambda \sum_{x \in \mathbb{Z}^d} p(x, n).$$

Now, we have to prove (3.9). In case $d = 1$ the proof of (3.9) is the same as that of (1.2). In order to prove (3.9) in case $d \geq 2$ we recall the following

THEOREM 3.4 ([44] Theorem 20.1) *Let $S(n)$ be a simple, symmetric random walk on $\mathbb{Z}^d$ and let $R(n)$ be the number of different vectors among $S(1), \ldots, S(n)$ i.e. $R(n)$ is the number of points visited by the particle during the first n steps. Then*

$$\mathbf{E}R(n) = \begin{cases} \frac{\pi n}{\log n} + O\left(\frac{n \log \log n}{(\log n)^2}\right) & \text{if } d = 2, \\ K_3 n + O(n^{1/2}) & \text{if } d = 3, \\ K_4 n + O(\log n) & \text{if } d = 4, \\ K_d n + \beta_d + O(n^{2-d/2}) & \text{if } d \geq 5 \end{cases}$$

where β_d and K_d are defined in Theorem 3.3.

Note that

$$\mathbf{E}R(n) = \sum_{x \in \mathbb{Z}^d} p(x, n).$$

(See [44] Lemma 20.1.) Hence we have (3.9) in case $d \geq 2$ as well.

Without having any new idea we get Theorems 1.4 - 1.12 and 2.1 - 2.3 for the model of this Section.

We are also interested to investigate how the initial distribution π_x of the present model can be changed. Let $\{\nu_x : x \in \mathbb{Z}^d\}$ be an array of i.i.d. non-negative integer valued r.v.'s. Hence ν_x defines a random field on $\mathbb{Z}^d$. Denote the points of this field by $\{Q_1, Q_2, \ldots\}$ in any order. Assume that there exists a one-to one mapping $f(\cdot)$ of the elements of the sequence $\{Q_1, Q_2, \ldots\}$ to the sequence $\{P_1, P_2, \ldots\}$ such that

$$\mathbf{P}\{\|f(Q_i) - P_i\| \geq \|P_i\|^{1/2+\varepsilon}\} < C_1 \exp\{-C_2 \|P_i\|^{\varepsilon}\} \tag{3.10}$$

for some $0 < \varepsilon < 1/2$, $C_1 > 0$, $C_2 > 0$.

Let $T_1(n), T_2(n), \ldots$ $(n = 0, 1, 2, \ldots)$ be a sequence of independent, simple, symmetric random walks on $\mathbb{Z}^d$ and define

$$Q_i(n) = Q_i + T_i(n) \qquad (i = 1, 2, \ldots;\ n = 0, 1, 2, \ldots).$$

By a simple modification of the coupling method of Section 3.2 we can prove that the above mentioned theorems remain true replacing the processes $\{P_i(n);\ i = 1, 2, \ldots;\ n = 0, 1, 2, \ldots\}$ by the processes $\{Q_i(n);\ i = 1, 2, \ldots\}$.

In order to show the meaning of condition (3.10) we give an

Example. Assume that

$$\mathbf{E}\nu_0 = \lambda \qquad \mathbf{E}\nu_0^3 < \infty.$$

Then it is easy to see that (3.10) is satisfied.

"I will indeed bless you,
and I will multiply your
descendants as the stars of
heaven and as the sand which
is on the seashore."

The First Book of Moses

Part II

BRANCHING RANDOM WALK

Chapter 4

Branching random walk starting with one particle

4.1 Branching process

At time $t = 0$ we have a particle. At time $t = 1$ this particle produces k ($k = 0, 1, 2, \ldots$) offsprings with probability p_k and dies. In case $k = 0$ we say that the process dies out at $t = 1$. In case $k > 0$ each offspring repeats this procedure independently. More formally speaking: let $\{Y(i,j),\ i = 1, 2, \ldots;\ j = 1, 2, \ldots\}$ be an array of i.i.d.r.v.'s with

$$\mathbf{P}\{Y(i,j) = k\} = p_k \qquad (k = 0, 1, 2, \ldots)$$

where

$$p_k \geq 0, \qquad \sum_{k=0}^{\infty} p_k = 1.$$

Define the sequence $\{B(n),\ n = 0, 1, 2, \ldots\}$ as follows

$$\begin{aligned}
B(0) &= 1, \\
B(1) &= Y(1,1), \\
B(2) &= Y(2,1) + Y(2,2) + \cdots + Y(2, B(1)), \\
\ldots & \quad \ldots\ldots\ldots \\
B(n) &= Y(n,1) + Y(n,2) + \cdots + Y(n, B(n-1)) \quad (n = 3, 4, \ldots).
\end{aligned}$$

Here $Y(i,j)$ can be considered as the number of offsprings of the j-th particle born at time $t = i$.

From now on we assume that

$$\mathbf{E}Y(i,j) = \sum_{k=0}^{\infty} kp_k = m < \infty$$

and

$$0 < \mathrm{Var}Y(i,j) = \sum_{k=0}^{\infty}(k-m)^2 p_k = \sigma^2 < \infty.$$

The branching process $\{B(n)\}$ is called subcritical if $m < 1$, critical if $m = 1$, supercritical if $m > 1$. The proof of the following wellknown theorem is very simple.

THEOREM 4.1

$$\mathbf{P}\{\lim_{n\to\infty} B(n) = 0\} = \begin{cases} 1 & \text{if} \quad m \le 1, \\ q & \text{if} \quad m > 1 \end{cases}$$

where $0 < q < 1$ *and in* $(0,1)$ q *is the unique solution of the equation*

$$g(q) = q, \qquad g(x) = \sum_{k=0}^{\infty} p_k x^k \qquad (0 < x < 1).$$

From now on up to the end of Chapter 6 we assume that

$$m > 1$$

and in this case we cite a more exact result.

THEOREM 4.2 ([28] Theorems 8.1, 8.3 and (8.4) pp. 13, 16). *There exists a nonnegative r. v.* B *such that*

$$\lim_{n\to\infty} m^{-n} B(n) = B \quad a.s., \qquad \mathbf{P}\{B = 0\} = q \tag{4.1}$$

$$\mathbf{E}\frac{B(n)}{m^n} = \mathbf{E}B = 1, \qquad \operatorname{Var} B = \frac{\sigma^2}{m^2 - m}, \tag{4.2}$$

$$\mathbf{E}\left|\frac{B(n)}{m^n} - B\right| = O(m^{-n/2}), \quad \mathbf{E}\left(\frac{B(n)}{m^n} - B\right)^2 = O(m^{-n}) \tag{4.3}$$

and the distribution of B *is absolutely continuous except for a jump of magnitude* q *at* 0.

4.2 The model

At time $t = 0$ a particle located in $0 \in \mathbb{Z}^d$ begins a random walk. It moves at time $t = 1$ with equal probabilities to one of the $2d$ neighbours of 0. Arriving to the new location it produces k offsprings with probability p_k $(k = 0, 1, 2, \ldots)$ and dies. Each of the offsprings move independently at time $t = 2$ to one of their neighbours. Arriving to the new locations each of them produce offsprings independently and they die. Repeating this procedure we obtain a *branching random walk*. Let $\lambda(x,t)$ $(x \in \mathbb{Z}^d,\ t = 0, 1, 2, \ldots)$ be the number of particles in x at t. We are interested in the limit properties of $\lambda(x,t)$ as $t \to \infty$.

At first we give a formal definition of $\lambda(x,t)$. Let $e_1, e_2, \ldots, e_d$ be the orthogonal unit vectors of $\mathbb{Z}^d$. Let

$$\{X(x,t,\mu), Z(x,t,\mu);\ x \in \mathbb{Z}^d,\ t = 0, 1, 2, \ldots,\ \mu = 1, 2, \ldots\}$$

be an array of independent r.v.'s with

$$\mathbf{P}\{X(x,t,\mu) = e_i\} = \mathbf{P}\{X(x,t,\mu) = -e_i\} = \frac{1}{2d}$$

$(i = 1, 2, \ldots, d;\ x \in \mathbb{Z}^d,\ t = 0, 1, 2, \ldots,\ \mu = 1, 2, \ldots)$,

$$\mathbf{P}\{Z(x, t, \mu) = k\} = p_k$$

$(k = 0, 1, 2, \ldots,\ x \in \mathbb{Z}^d,\ t = 0, 1, 2, \ldots,\ \mu = 1, 2, \ldots)$, where

$$p_k \geq 0, \qquad \sum_{k=0}^{\infty} p_k = 1, \qquad \sum_{k=0}^{\infty} k p_k = m > 1,$$

$$\sum_{k=0}^{\infty} (k - m)^2 p_k = \sigma^2 < \infty.$$

Further put

$$I_u(v) = \begin{cases} 1 & \text{if} \quad u = v, \\ 0 & \text{if} \quad u \neq v, \end{cases}$$
$$C(x, R) = \{y : \ y \in \mathbb{Z}^d,\ |y - x| \leq R\},$$
$$C(x) = \{y : \ y \in \mathbb{Z}^d,\ |y - x| = 1\}.$$

Then we give the definition of the array

$$\{\lambda(x, t);\ x \in \mathbb{Z}^d,\ t = 0, 1, 2, \ldots\}$$

as follows. Let

$$\lambda(x, 0) = \begin{cases} 1 & \text{if} \quad x = 0, \\ 0 & \text{if} \quad x \neq 0 \end{cases}$$

and

$$\lambda(x, t) = \sum_{y \in C(x)} \sum_{\mu=1}^{\lambda(y, t-1)} I_{x-y}(X(y, t-1, \mu)) Z(y, t-1, \mu)$$

if $x \in \mathbb{Z}^d$ and $t = 1, 2, \ldots$.

The intuitive meaning of the above definition is the following. In order to get the number of the particles located in x at time t, consider the particles located in one of the neighbours of x at time $t - 1$. The μ-th particle located in $y \in C(x)$ at time $t - 1$ is moving to x if $X(y, t-1, \mu) = x - y$ i.e. if $I_{x-y}(X(y, t-1, \mu)) = 1$. If this particle moves to x then it will produce there

$$Z(y, t-1, \mu) = I_{x-y}(X(y, t-1, \mu)) Z(y, t-1, \mu)$$

offsprings.

Fig. 1 shows the branching random walk in case $d = 1$, $t = 0, 1, 2, 3, 4, 5$,

$$X(0,0,1) = 1,$$
$$Z(0,0,1) = 3,$$

$$X(1,1,1) = X(1,1,2) = -1, \ X(1,1,3) = 1,$$
$$Z(1,1,1) = Z(1,1,2) = 2, \ Z(1,1,3) = 3,$$

$$X(0,2,1) = X(0,2,2) = X(0,2,3) = 1, \ X(0,2,4) = -1,$$
$$Z(0,2,1) = 3, \ Z(0,2,2) = 1, \ Z(0,2,3) = 0, \ Z(0,2,4) = 1,$$

$$X(2,2,1) = X(2,2,2) = 1, \ X(2,2,3) = -1,$$
$$Z(2,2,1) = 2, \ Z(2,2,2) = 1, \ Z(2,2,3) = 1,$$

$$X(-1,3,1) = -1,$$
$$Z(-1,3,1) = 0,$$

$$X(1,3,1) = X(1,3,2) = X(1,3,3) = -1, \ X(1,3,4) = X(1,3,5) = 1,$$
$$Z(1,3,1) = 2, \ Z(1,3,2) = 1, \ Z(1,3,3) = 0, \ Z(1,3,4) = 2, \ Z(1,3,5) = 1,$$

$$X(3,3,1) = X(3,3,2) = -1, \ X(3,3,3) = 1,$$
$$Z(3,3,1) = 1, \ Z(3,3,2) = 0, \ Z(3,3,3) = 4,$$

$$X(0,4,1) = X(0,4,2) = -1, \ X(0,4,3) = 1,$$
$$X(2,4,1) = X(2,4,2) = X(2,4,3) = 1, \ X(2,4,4) = -1,$$
$$X(4,4,1) = X(4,4,2) = -1, \ X(4,4,3) = X(4,4,4) = 1.$$

The numbers in the circles give the actual values of $\lambda(x,t)$.

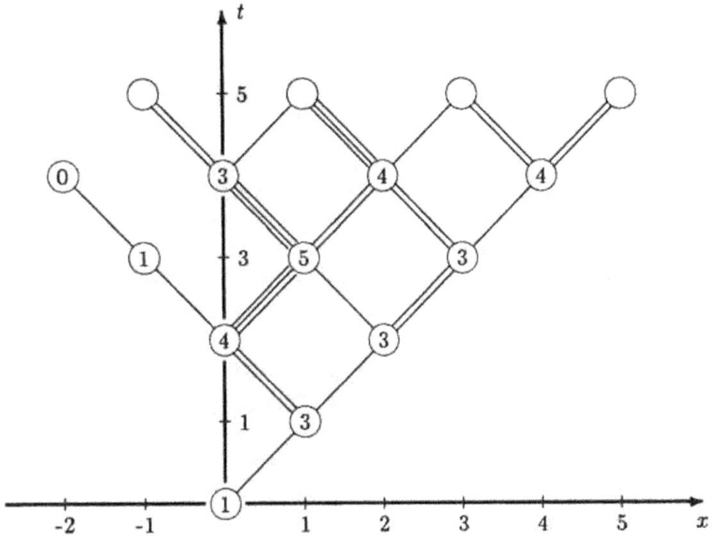

Fig. 1

Hence

$$\lambda(x,0) = \begin{cases} 1 & \text{if} \quad x = 0, \\ 0 & \text{otherwise}, \end{cases}$$

$$\lambda(x,1) = \begin{cases} 3 & \text{if} \quad x = 1, \\ 0 & \text{otherwise}, \end{cases}$$

$$\lambda(x,2) = \begin{cases} 4 & \text{if} \quad x = 0, \\ 3 & \text{if} \quad x = 2, \\ 0 & \text{otherwise}, \end{cases}$$

$$\lambda(x,3) = \begin{cases} 1 & \text{if} \quad x = -1, \\ 5 & \text{if} \quad x = 1, \\ 3 & \text{if} \quad x = 3, \\ 0 & \text{otherwise}, \end{cases}$$

$$\lambda(x,4) = \begin{cases} 3 & \text{if} \quad x = 0, \\ 4 & \text{if} \quad x = 2, \\ 4 & \text{if} \quad x = 4, \\ 0 & \text{otherwise}, \end{cases}$$

and $B(0) = 1,\ B(1) = 3,\ B(2) = 7,\ B(3) = 9,\ B(4) = 11.$

4.3 On the moments of $\lambda(x,t)$

Introduce the following notations:

(i) let $\{S_t;\ t = 0,1,2,\ldots\}$ be a simple, symmetric, nearest neighbour random walk on $\mathbb{Z}^d$ with $S_0 = 0$,

(ii)
$$p(u \rightsquigarrow v, t) = \mathbf{P}\{S_{s+t} = v \mid S_s = u\},$$

(iii)
$$f(x,T,t) = m^{T-t} \sum_{y\in\mathbb{Z}^d} \lambda(y,t)p(y \rightsquigarrow x, T-t) \quad (0 \le t \le T),$$

(iv) let
$$\mathcal{F}(T) = \mathcal{F}\{\lambda(x,t);\ x \in \mathbb{Z}^d,\ t = 0,1,2,\ldots,T\}$$
be the smallest σ-algebra with respect to which the array
$$\{\lambda(x,t);\ x \in \mathbb{Z}^d,\ t = 0,1,2,\ldots,T\}$$
is measurable,

(v) let
$$\mathcal{G}(T) = \mathcal{G}\{X(y,s,\mu), Z(y,s,\mu);\ s \ge T,\ y \in \mathbb{Z}^d,\ \mu \ge 1\}$$
be the smallest σ-algebra with respect to which the r.v.'s in the brackets are measurable,

(vi)
$$\sum_{y\in\mathbb{Z}^d} \lambda(y,t) = B(t).$$

Note that $\mathcal{F}(T)$ and $\mathcal{G}(T)$ are independent σ-algebras. Further by the definition of $\lambda(x,T)$ we have

$$\mathbf{E}(\lambda(x,T) \mid \mathcal{F}(T-1)) = \frac{m}{2d} \sum_{y\in C(x)} \lambda(y, T-1).$$

At first we present a number of lemmas.

LEMMA 4.1 ([32] Theorem 1.2.1) *Assume that*

$$x = (x_1, x_2, \ldots, x_d) \equiv t \pmod 2$$

i.e.

$$x_1 + x_2 + \cdots + x_d \equiv t \pmod 2.$$

Then

$$\left| p(0 \rightsquigarrow x, t) - 2\left(\frac{d}{2\pi t}\right)^{d/2} \exp\left(-\frac{d\|x\|^2}{2t}\right)\right| \le O(t^{-(d+2)/2})$$

as $t \to \infty$.

LEMMA 4.2

$$\sum_{x\in\mathbb{Z}^d} p^2(0 \rightsquigarrow x,t) = p(0 \rightsquigarrow 0,2t) \sim 2\left(\frac{d}{4\pi t}\right)^{d/2} \qquad (t = 1,2,\ldots).$$

Proof.

$$\sum_{x\in\mathbb{Z}^d} p^2(0 \rightsquigarrow x,t) = \sum_{x\in\mathbb{Z}^d} p(0 \rightsquigarrow x,t)p(x \rightsquigarrow 0,t) = p(0 \rightsquigarrow 0,2t).$$

The asymptotic relation follows from Lemma 4.1.

LEMMA 4.3 *For any $0 \le t \le T$ and $x \in \mathbb{Z}^d$ we have*

$$\mathbf{E}(\lambda(x,T) \mid \mathcal{F}(t)) = f(x,T,t). \tag{4.4}$$

Proof. Since

$$\mathbf{E}(\lambda(x,t) \mid \mathcal{F}(t)) = \lambda(x,t)$$

and

$$f(x,T,T) = \sum_{y\in\mathbb{Z}^d} \lambda(y,T)p(y \rightsquigarrow x,0) = \lambda(x,T),$$

we have (4.4) in case $T = t$. It is also easy to see that (4.4) holds if $T = 1$ and $t = 0$. Now we give the proof by induction. Assume that (4.4) holds true for T, for any $0 \le t \le T$ and $x \in \mathbb{Z}^d$. Then

$$\begin{aligned}
&\mathbf{E}(\lambda(x,T+1) \mid \mathcal{F}(t)) = \\
&= \mathbf{E}\left(\sum_{y\in C(x)} \sum_{\mu=1}^{\lambda(y,T)} I_{x-y}(X(y,T,\mu))Z(y,T,\mu) \,\middle|\, \mathcal{F}(t)\right) = \\
&= \frac{m}{2d}\sum_{y\in C(x)} \mathbf{E}(\lambda(y,T) \mid \mathcal{F}(t)) = \frac{m}{2d}\sum_{y\in C(x)} f(y,T,t) = \\
&= \frac{m}{2d}\sum_{y\in C(x)} m^{T-t}\sum_{z\in\mathbb{Z}^d} \lambda(z,t)p(z \rightsquigarrow y,T-t) = \\
&= m^{T-t+1}\sum_{z\in\mathbb{Z}^d} \lambda(z,t)\frac{1}{2d}\sum_{y\in C(x)} p(z \rightsquigarrow y,T-t) = \\
&= m^{T-t+1}\sum_{z\in\mathbb{Z}^d} \lambda(z,t)p(z \rightsquigarrow x,T-t+1) = f(x,T+1,t).
\end{aligned}$$

Hence we have (4.4).

Note that (4.4) can be interpreted as follows: $f(x,T,t)$ is an unbiased predictor of $\lambda(x,T)$ observing the process up to t.

LEMMA 4.4

$$f(x,T,t) =$$
$$= m^{T-t} \sum_{y\in \mathbb{Z}^d} \sum_{\mu=1}^{\lambda(y,t-1)} Z(y,t-1,\mu) \sum_{u\in C(y)} I_{u-y}(X(y,t-1,\mu))p(u \rightsquigarrow x, T-t). \quad (4.5)$$

Proof. Since

$$I_{u-y}(X(y,t-1,\mu)) = 0 \quad \text{if} \quad u \notin C(y),$$

we have

$$f(x,T,t) =$$
$$= m^{T-t} \sum_{y\in \mathbb{Z}^d} \sum_{u\in C(y)} \sum_{\mu=1}^{\lambda(u,t-1)} I_{y-u}(X(u,t-1,\mu))Z(u,t-1,\mu)p(y \rightsquigarrow x, T-t) =$$
$$= m^{T-t} \sum_{y\in \mathbb{Z}^d} \sum_{u\in \mathbb{Z}^d} \sum_{\mu=1}^{\lambda(u,t-1)} I_{y-u}(X(u,t-1,\mu))Z(u,t-1,\mu)p(y \rightsquigarrow x, T-t) =$$
$$= m^{T-t} \sum_{u\in \mathbb{Z}^d} \sum_{\mu=1}^{\lambda(u,t-1)} Z(u,t-1,\mu) \sum_{y\in \mathbb{Z}^d} I_{y-u}(X(u,t-1,\mu))p(y \rightsquigarrow x, T-t).$$

Hence we have (4.5).

LEMMA 4.5

$$\mathbf{E}\lambda(x,T) = m^T p(0 \rightsquigarrow x, T), \quad (4.6)$$
$$\mathbf{E}(f(x,T,t) \mid \mathcal{F}(t-1)) = f(x,T,t-1),$$
$$= m^{T-t+1} \sum_{y\in \mathbb{Z}^d} \lambda(y,t-1)p(y \rightsquigarrow x, T-t+1), \quad (4.7)$$
$$\mathbf{E}f(x,T,t) = m^T p(0 \rightsquigarrow x, T) \quad (t = 0,1,2,\ldots,T). \quad (4.8)$$

Proof. (4.4) with $t = 0$ implies

$$\mathbf{E}\lambda(x,T) = \mathbf{E}(\lambda(x,T) \mid \mathcal{F}(0)) = f(x,T,0) =$$
$$= m^T \sum_{y\in \mathbb{Z}^d} \lambda(y,0)p(y \rightsquigarrow x, T) = m^T p(0 \rightsquigarrow x, T).$$

Hence we have (4.6). By (4.4) we have

$$\mathbf{E}(f(x,T,t) \mid \mathcal{F}(t-1)) = \mathbf{E}(\mathbf{E}(\lambda(x,T) \mid \mathcal{F}(t)) \mid \mathcal{F}(t-1)) =$$
$$= \mathbf{E}(\lambda(x,T) \mid \mathcal{F}(t-1)) = f(x,T,t-1).$$

Hence we have (4.7).

By (4.4) and (4.6)

$$\mathbf{E}f(x,T,t) = \mathbf{E}\lambda(x,T) = m^T p(0 \rightsquigarrow x, T)$$

and we have (4.8).

LEMMA 4.6

$$\sum_{x\in\mathbb{Z}^d} \mathbf{E}((f(x,T,t)-f(x,T,t-1))^2 \mid \mathcal{F}(t-1)) =$$
$$= m^{2T-2t}(B(t-1)((m^2+\sigma^2)p(0\rightsquigarrow 0,2T-2t)-m^2p(0\rightsquigarrow 0,2T-2t+2))) \sim$$
$$\sim 2\sigma^2\left(\frac{d}{4\pi}\right)^{d/2} m^{2T-2t}B(t-1)\frac{1}{(T-t)^{d/2}}$$

where the asymptotic relation holds true if $T-t\to\infty$.

Proof. By (4.5) we have

$$f(x,T,t)-f(x,T,t-1) =$$
$$= m^{T-t}\sum_{y\in\mathbb{Z}^d}\sum_{\mu=1}^{\lambda(y,t-1)} Z(y,t-1,\mu)\sum_{u\in C(y)} I_{u-y}(X(y,t-1,\mu))p(u\rightsquigarrow x,T-t)-$$
$$-m^{T-t+1}\sum_{y\in\mathbb{Z}^d}\lambda(y,t-1)p(y\rightsquigarrow x,T-t+1) =$$
$$= m^{T-t}\sum_{y\in\mathbb{Z}^d}\sum_{\mu=1}^{\lambda(y,t-1)} Q(y,\mu)$$

where

$$Q(y,\mu) = Z(y,t-1,\mu)\sum_{u\in C(y)} I_{u-y}(X(y,t-1,\mu))p(u\rightsquigarrow x,T-t)-$$
$$-mp(y\rightsquigarrow x,T-t+1).$$

Observe that

$$\mathbf{E}\left(Z(y,t-1,\mu)\sum_{u\in C(y)} I_{u-y}(X(y,t-1,\mu))p(u\rightsquigarrow x,T-t)\right) =$$
$$=\frac{m}{2d}\sum_{u\in C(y)} p(u\rightsquigarrow x,T-t) = mp(y\rightsquigarrow x,T-t+1).$$

Hence

$$\mathbf{E}(Q^2(y,\mu)\mid\mathcal{F}(t-1)) = \mathbf{E}Q^2(y,\mu) = (m^2+\sigma^2)\frac{1}{2d}\sum_{u\in C(y)} p^2(u\rightsquigarrow x,T-t)-$$
$$-m^2p^2(y\rightsquigarrow x,T-t+1)$$

and

$$\mathbf{E}((f(x,T,t)-f(x,T,t-1))^2\mid\mathcal{F}(t-1)) = m^{2T-2t}\sum_{y\in\mathbb{Z}^d}\lambda(y,t-1)\Delta$$

where

$$\Delta = \Delta(x,y,t,T) = \frac{m^2+\sigma^2}{2d} \sum_{u\in C(y)} p^2(u \rightsquigarrow x, T-t) - m^2p^2(y \rightsquigarrow x, T-t+1).$$

Consequently by Lemma 4.2

$$m^{2t-2T} \sum_{x\in\mathbb{Z}^d} \mathbf{E}((f(x,T,t) - f(x,T,t-1))^2 \mid \mathcal{F}(t-1)) = \sum_{y\in\mathbb{Z}^d} \lambda(y,t-1) \sum_{x\in\mathbb{Z}^d} \Delta =$$

$$= \sum_{y\in\mathbb{Z}^d} \lambda(y,t-1)((m^2+\sigma^2)p(0 \rightsquigarrow 0, 2T-2t) - m^2p(0 \rightsquigarrow 0, 2T-2t+2)) =$$

$$= B(t-1)((m^2+\sigma^2)p(0 \rightsquigarrow 0, 2T-2t) - m^2p(0 \rightsquigarrow 0, 2T-2t+2)).$$

Hence we have Lemma 4.6 by Lemma 4.1.

LEMMA 4.7 *Assume that $T-t \to \infty$. Then*

$$\sum_{x\in\mathbb{Z}^d} \mathbf{E}((\lambda(x,T) - f(x,T,t))^2 \mid \mathcal{F}(t)) \sim 2\sigma^2 \left(\frac{d}{4\pi}\right)^{d/2} B(t) \sum_{i=1}^{T-t} i^{-d/2} m^{T+i-t-2}$$

Proof. Clearly by (4.4) we have

$$\begin{aligned}
&\mathbf{E}((\lambda(x,T) - f(x,T,t))^2 \mid \mathcal{F}(t)) = \\
&= \mathbf{E}(\mathbf{E}((\lambda(x,T) - f(x,T,t))^2 \mid \mathcal{F}(T-1)) \mid \mathcal{F}(t)) = \\
&= \mathbf{E}(\mathbf{E}((\lambda(x,T) - f(x,T,T-1))^2 \mid \mathcal{F}(T-1)) \mid \mathcal{F}(t)) + \\
&\quad + \mathbf{E}((f(x,T,T-1) - f(x,T,t))^2 \mid \mathcal{F}(t)).
\end{aligned}$$

Since $\lambda(x,T) = f(x,T,T)$ by Lemma 4.6 with $t=T$ we have

$$\begin{aligned}
&\sum_{x\in\mathbb{Z}^d} \mathbf{E}((\lambda(x,T) - f(x,T,t))^2 \mid \mathcal{F}(t)) = \\
&= ((m^2+\sigma^2)p(0 \rightsquigarrow 0,0) - m^2p(0 \rightsquigarrow 0,2))m^{T-t-1}B(t) + \\
&\quad + \sum_{x\in\mathbb{Z}^d} \mathbf{E}((f(x,T,T-1) - f(x,T,t))^2 \mid \mathcal{F}(t)).
\end{aligned}$$

Similarly we get

$$\begin{aligned}
&\sum_{x\in\mathbb{Z}^d} \mathbf{E}((f(x,T,T-1) - f(x,T,t))^2 \mid \mathcal{F}(t)) = \\
&= \sum_{x\in\mathbb{Z}^d} \mathbf{E}(\mathbf{E}((f(x,T,T-1) - f(x,T,t))^2 \mid \mathcal{F}(T-2)) \mid \mathcal{F}(t)) = \\
&= \sum_{x\in\mathbb{Z}^d} \mathbf{E}(\mathbf{E}((f(x,T,T-1) - f(x,T,T-2))^2 \mid \mathcal{F}(T-2)) \mid \mathcal{F}(t)) + \\
&\quad + \sum_{x\in\mathbb{Z}^d} \mathbf{E}((f(x,T,T-2) - f(x,T,t))^2 \mid \mathcal{F}(t)) = \\
&= m^{T-t}B(t)((m^2+\sigma^2)p(0 \rightsquigarrow 0,2) - m^2p(0 \rightsquigarrow 0,4)) + \\
&\quad + \sum_{x\in\mathbb{Z}^d} \mathbf{E}((f(x,T,T-2) - f(x,T,t))^2 \mid \mathcal{F}(t)).
\end{aligned}$$

Going step by step by Lemma 4.1 we obtain

$$\sum_{x\in\mathbb{Z}^d} \mathbf{E}((\lambda(x,T)-f(x,T,t))^2 \mid \mathcal{F}(t)) =$$

$$= m^{T-t-1}\sum_{i=0}^{T-t-1} m^i((m^2+\sigma^2)p(0\rightsquigarrow 0,2i)-m^2p(0\rightsquigarrow 0,2i+2))B(t)\sim$$

$$\sim m^{T-t-1}B(t)\sum_{i=1}^{T-t}\sigma^2 m^{i-1}2\left(\frac{d}{4\pi i}\right)^{d/2} = m^{T-t-2}\sigma^2 2\left(\frac{d}{4\pi}\right)^{d/2} B(t)\sum_{i=1}^{T-t} m^i\left(\frac{1}{i}\right)^{d/2} =$$

$$= 2\sigma^2\left(\frac{d}{4\pi}\right)^{d/2} B(t)\sum_{i=1}^{T-t} m^{T+i-t-2}\frac{1}{i^{d/2}}.$$

Hence we have Lemma 4.7.

LEMMA 4.8

$$\mathbf{E}\left(\sum_{x\in\mathbb{Z}^d}\left(\frac{\lambda(x,T)}{m^T}-\frac{f(x,T,t)}{m^T}\right)^2 \,\middle|\, \mathcal{F}(t)\right) \le C\frac{1}{m^t(T-t)^{d/2}}\frac{B(t)}{m^t}$$

with an absolute constant $C>0$.

Proof. Lemma 4.8 is a trivial consequence of Lemma 4.7.

THEOREM 4.3 *For any $0\le\alpha\le 1$ we have*

$$\mathbf{E}\left(\sum_{x\in C(0,T^\alpha)}\left|\frac{\lambda(x,T)}{m^T}-\frac{f(x,T,t)}{m^T}\right| \,\middle|\, \mathcal{F}(t)\right) \le C(2T^\alpha)^{d/2}\left(\frac{1}{m^t(T-t)^{d/2}}\frac{B(t)}{m^t}\right)^{1/2} \tag{4.9}$$

and

$$\mathbf{E}\sum_{x\in C(0,T^\alpha)}\left|\frac{\lambda(x,T)}{m^T}-\frac{f(x,T,t)}{m^T}\right| \le C(2T^\alpha)^{d/2}(m^t(T-t)^{d/2})^{-1/2} \tag{4.10}$$

with an absolute constant $C>0$.

Proof. By Lemma 4.8 and the Cauchy inequality we have

$$\mathbf{E}\left(\sum_{x\in C(0,T^\alpha)}\left|\frac{\lambda(x,T)}{m^T}-\frac{f(x,T,t)}{m^T}\right| \,\middle|\, \mathcal{F}(t)\right)\le$$

$$\le (2T^\alpha)^{d/2}\left(\mathbf{E}\left(\sum_{x\in C(0,T^\alpha)}\left(\frac{\lambda(x,T)}{m^T}-\frac{f(x,T,t)}{m^T}\right)^2 \,\middle|\, \mathcal{F}(t)\right)\right)^{1/2}\le$$

$$\le C(2T^\alpha)^{d/2}\left(\frac{1}{m^t(T-t)^{d/2}}\frac{B(t)}{m^t}\right)^{1/2}.$$

Hence we have (4.9). (4.2) and the Cauchy inequality implies $\mathbf{E}(m^{-t}B_t)^{1/2}\le 1$ which, in turn, by (4.9) implies (4.10).

4.4 Global limit theorems

The results of Section 4.3 especially (4.6), (4.8) and Theorem 4.3 suggest that $m^{-T}\lambda(x,T)$ should behave like $p(0 \rightsquigarrow x, T)$.

In fact, Theorem 4.3 told us that $f(x,T,t)$ is really a good estimator of $\lambda(x,T)$ if t is big enough. Hence we have to show that $m^{-T}f(x,T,t)$ behaves like $p(0 \rightsquigarrow x,T)B$. Observing the process up to t we get $B(t)$ as well as the distribution of the $B(t)$ particles in $\mathbb{Z}^d$. We will show that the distribution of these $B(t)$ particles does not have a significant influence to $\lambda(x,t)$ if t is small, however the value of $B(t)$ does. This picture suggests that we have to choose t big enough in order to get $\lambda(x,T)$ close to $f(x,T,t)$ and to choose t small enough to minimize the influence of the distribution of the $B(t)$ particles.

In this and the next Sections we execute these estimations. Our main result runs as follows.

THEOREM 4.4 *For any* $0 < \varepsilon < 1/2$ *there exists a* $C = C(\varepsilon) > 0$ *such that for any* $T = 1, 2, \ldots$ *we have*

$$\mathbf{E}\sum_{x\in\mathbb{Z}^d}\left|\frac{\lambda(x,T)}{m^T} - p(0 \rightsquigarrow x,T)B\right| \leq CT^{-(1/2-\varepsilon)}$$

where (cf. Theorem 4.2)

$$B = \lim_{T\to\infty} m^{-T}B(T)$$

and

$$B(T) = \sum_{x\in\mathbb{Z}^d}\lambda(x,T).$$

In order to prove Theorem 4.4 we prove two lemmas.

LEMMA 4.9 *Let* $1 \leq t \leq T$, $x \equiv T \pmod 2$ *and*

$$D(T) = \{x:\ x\in C(0,T),\ x\equiv T \pmod 2\}.$$

Then

$$\mathbf{E}(\lambda(x,T) \mid \mathcal{F}(t)) \geq \inf_{y\in D(t)} \mathbf{E}(\lambda(x,T) \mid \lambda(y,t) = B(t)) = m^T\frac{B(t)}{m^t}\inf_{y\in D(t)} p(0 \rightsquigarrow x-y, T-t)$$

and

$$\mathbf{E}(\lambda(x,T) \mid \mathcal{F}(t)) \leq \sup_{y\in D(t)} \mathbf{E}(\lambda(x,T) \mid \lambda(y,t) = B(t)) = m^T\frac{B(t)}{m^t}\sup_{y\in D(t)} p(0 \rightsquigarrow x-y, T-t).$$

Proof. Observing that $\lambda(y,t) = 0$ if $y \notin D(t)$ we have the inequalities of the above two relations. The equalities in them follow from (4.4).

LEMMA 4.10 *Assume the conditions of Lemma 4.9 and let* $|x|t = o(T)$ *and* $t^2 = o(T)$ *as* $T \to \infty$. *Then we have*

$$\frac{p(0 \leadsto x, T)}{\inf_{y \in D(t)} p(0 \leadsto x - y, T - t)} \leq 1 + O\left(\frac{|x|t + t^2}{T}\right) \tag{4.11}$$

and

$$\frac{p(0 \leadsto x, T)}{\sup_{y \in D(t)} p(0 \leadsto x - y, T - t)} \geq 1 - O\left(\frac{|x|t + t^2}{T}\right). \tag{4.12}$$

Further if $1/2 < \alpha < 1$ *then*

$$\sum_{x \in D(T) - D(T^\alpha)} p(0 \leadsto x, T) = \exp(-O(T^{2\alpha - 1})). \tag{4.13}$$

Proof. At first we consider the case $d = 1$. Then

$$p(0 \leadsto 2x, 2T) = \binom{2T}{T - x} \frac{1}{2^{2T}}$$

and by the Stirling formula we get

$$\frac{p(0 \leadsto 2x, 2T)}{p(0 \leadsto 2x, 2T - 2t)} = 1 + O\left(\frac{|x|t + t^2}{T}\right) \tag{4.14}$$

provided that

$$|x|t = o(T) \quad \text{and} \quad t^2 = o(T) \quad \text{as} \quad T \to \infty.$$

Similarly

$$\frac{p(0 \leadsto 2x, 2T)}{p(0 \leadsto 2z, 2T)} = 1 + O\left(\frac{x^2 - z^2}{T}\right) \tag{4.15}$$

provided that

$$x^2 - z^2 = o(T), \qquad z = o(T) \qquad \text{and} \qquad x = o(T).$$

Consequently

$$\frac{p(0 \leadsto 2x, 2T)}{p(0 \leadsto 2z, 2T - 2t)} = 1 + O\left(\frac{|x|t + t^2 + x^2 - z^2}{T}\right) \tag{4.16}$$

provided that

$$|x|t = o(T), \quad z = o(T), \quad x^2 - z^2 = o(T) \text{ and } t^2 = o(T) \text{ as } T \to \infty.$$

Since $|x^2-(x-y)^2| \le 2|x|t+t^2$ if $y \in D(t)$, we have (4.11) and (4.12) in case $d=1$. In case $d \ge 2$ let

$$T_i = \#\{j:\ 0 \le j \le T-1,\ |(e_i, S_{j+1}-S_j)| = 1\}$$

i.e. T_i is the number of those steps in $[0,T]$ when the particle moves in the direction e_i or $-e_i$. Then for any $0<\varepsilon<1$ we have

$$\begin{aligned} p(0 \rightsquigarrow x, T) &= \\ &= \sum\nolimits^{*} \mathbf{P}\left\{\bigcap_{i=1}^{d}\{S_{t_i}^{(i)} = x_i\} \mid \bigcap_{i=1}^{d}\{T_i = t_i\}\right\} \mathbf{P}\left\{\bigcap_{i=1}^{d}\{T_i = t_i\}\right\} = \\ &= \sum\nolimits^{*} \left(\prod_{i=1}^{d} \mathbf{P}\{S_{t_i}^{(i)} = x_i\}\right) \mathbf{P}\left\{\bigcap_{i=1}^{d}\{T_i = t_i\}\right\} = \\ &= (1+o(1)) \sum\nolimits_{\varepsilon} \left(\prod_{i=1}^{d} \mathbf{P}\{S_{t_i}^{(i)} = x_i\}\right) \mathbf{P}\left\{\bigcap_{i=1}^{d}\{T_i = t_i\}\right\} \end{aligned} \tag{4.17}$$

where

$$S_t = (S_t^{(1)}, S_t^{(2)}, \ldots, S_t^{(d)}),$$

$$x = (x_1, x_2, \ldots, x_d),$$

$\sum^{*}$ is the summation extended for all d-tuples $(t_1, t_2, \ldots, t_d)$ for which $t_1 + t_2 + \cdots + t_d = T$ and $t_i \equiv x_i \pmod 2$,

$\sum_{\varepsilon}$ is the summation extended for all d-tuples $(t_1, t_2, \ldots, t_d)$ for which $t_1 + t_2 + \ldots + t_d = T$, $t_i \equiv x_i \pmod 2$ and $|t_i/T - 1/d| \le \varepsilon$ $(i = 1, 2, \ldots, d)$.

(4.17) and (4.16) combined imply (4.11) and (4.12). (4.13) follows from Lemma 4.1.

Proof of Theorem 4.4. For any $0 \le t \le T$ and $0 < \alpha < 1$ we have

$$\begin{aligned} \sum_{x\in D(T)} \left|\frac{\lambda(x,T)}{m^T} - p(0 \rightsquigarrow x, T)B\right| &\le \\ \le \sum_{x\in D(T)-D(T^\alpha)} \frac{\lambda(x,T)}{m^T} &+ B \sum_{x\in D(T)-D(T^\alpha)} p(0 \rightsquigarrow x, T) + \\ + \sum_{x\in D(T^\alpha)} &\left|\frac{\lambda(x,T)}{m^T} - \frac{\mathbf{E}(\lambda(x,T) \mid \mathcal{F}(t))}{m^T}\right| + \\ + \sum_{x\in D(T^\alpha)} &\left|\frac{\mathbf{E}(\lambda(x,T) \mid \mathcal{F}(t))}{m^T} - p(0 \rightsquigarrow x, T)\frac{B(t)}{m^t}\right| + \\ + \sum_{x\in D(T^\alpha)} &\left|p(0 \rightsquigarrow x, T)\frac{B(t)}{m^t} - p(0 \rightsquigarrow x, T)B\right|. \end{aligned} \tag{4.18}$$

Choosing $\alpha > 1/2$ by (4.13) and (4.6)

$$\sum_{x \in D(T)-D(T^\alpha)} p(0 \rightsquigarrow x, T) \leq \exp(-O(T^{2\alpha-1})), \tag{4.19}$$

$$\sum_{x \in D(T)-D(T^\alpha)} \mathbf{E}\left(\frac{\lambda(x,T)}{m^T}\right) \leq \exp(-O(T^{2\alpha-1})). \tag{4.20}$$

Let $T = [t^\beta]$ with $\beta > 1$ by Theorem 4.3 and (4.4) we have

$$\sum_{x \in D(T^\alpha)} \mathbf{E}\left|\frac{\lambda(x,T)}{m^T} - \frac{\mathbf{E}(\lambda(x,T) \mid \mathcal{F}(t))}{m^T}\right| \leq$$

$$\leq C(2T^\alpha)^{d/2}(m^t(T-t)^{d/2})^{-1/2} \leq \exp(-O(T^{1/\beta})). \tag{4.21}$$

Choosing $\alpha > 1/2$ close enough to $1/2$ and β big enough by Lemmas 4.9 and 4.10 we have

$$\sum_{x \in D(T^\alpha)} \left|\frac{\mathbf{E}(\lambda(x,T) \mid \mathcal{F}(t))}{m^T} - p(0 \rightsquigarrow x, T)\frac{B(t)}{m^t}\right| \leq O(T^{-(1/2-\varepsilon)}) \tag{4.22}$$

for any $\varepsilon > 0$. By (4.3)

$$\sum_{x \in D(T^\alpha)} \mathbf{E}\left(\left|p(0 \rightsquigarrow x, T)\frac{B(t)}{m^t} - p(0 \rightsquigarrow x, T)B\right|\right) \leq \exp(-O(T^{1/\beta})). \tag{4.23}$$

By inequalities (4.18)–(4.23) we have Theorem 4.4.

A strong version of Theorem 4.4 can be obtained similarly. In fact we prove

THEOREM 4.5 *For any $\varepsilon > 0$*

$$\lim_{T\to\infty} T^{1/2-\varepsilon} \sum_{x \in \mathbb{Z}^d} \left|\frac{\lambda(x,T)}{m^T} - p(0 \rightsquigarrow x, T)B\right| = 0 \quad a.s.$$

Proof. By (4.20), (4.21) and (4.23) respectively for any $K > 0$ we have

$$\lim_{T\to\infty} T^K \sum_{x \in D(T)-D(T^\alpha)} \frac{\lambda(x,T)}{m^T} = 0 \qquad \text{a.s.} \tag{4.24}$$

resp.

$$\lim_{T\to\infty} T^K \sum_{x \in D(T^\alpha)} \left|\frac{\lambda(x,T)}{m^T} - \frac{\mathbf{E}(\lambda(x,T) \mid \mathcal{F}(t))}{m^T}\right| = 0 \qquad \text{a.s.} \tag{4.25}$$

$$\lim_{T\to\infty} T^K \sum_{x \in D(T^\alpha)} \left|p(0 \rightsquigarrow x, T)\frac{B(T)}{m^T} - p(0 \rightsquigarrow x, T)B\right| = 0 \qquad \text{a.s.} \tag{4.26}$$

respectively.

Hence we have Theorem 4.5 by (4.18), (4.19), (4.24), (4.25), (4.22) and (4.26).

Theorem 4.5 easily implies

THEOREM 4.6 *For any $x \in \mathbb{R}^d$ and $\varepsilon > 0$ we have*

$$\lim_{T\to\infty} T^{1/2-\varepsilon}\left|\sum_{y\le xT^{1/2}} \frac{\lambda(y,T)}{m^T} - B\Phi(x)\right| = 0 \qquad a.s.$$

where

$$\Phi(x) = \Phi(x_1,x_2,\ldots,x_d) = \\ = \frac{1}{(2\pi)^{d/2}}\int_{-\infty}^{x_1}\cdots\int_{-\infty}^{x_d}\exp\left(-\frac{y_1^2+y_2^2+\cdots+y_d^2}{2}\right)dy_1dy_2\ldots dy_d.$$

4.5 Local limit theorems

It looks likely that if in Theorems 4.4 and 4.5 the sums are taken for a small set then the rate of convergence is better. In fact we prove

THEOREM 4.7 *For any $0 \le \gamma \le 1/2$, $\varepsilon > 0$ and $0 \le a < \infty$ we have*

$$\mathbf{E}\left(\sum_{x\in D(aT^\gamma)}\left|\frac{\lambda(x,T)}{m^T} - p(0\rightsquigarrow x,T)B\right|\right) \le CT^{-(d+2-2\gamma(d+1)-\varepsilon)/2}, \tag{4.27}$$

$$\lim_{T\to\infty} T^{(d+2-2\gamma(d+1)-\varepsilon)/2}\sum_{x\in D(aT^\gamma)}\left|\frac{\lambda(x,T)}{m^T} - p(0\rightsquigarrow x,T)B\right| = 0 \quad a.s. \tag{4.28}$$

THEOREM 4.8 *Let $x = x(T) \in D(T^\gamma)$ $(0 \le \gamma \le 1)$ be a sequence of vectors. Then for any $0 < \varepsilon < \gamma$*

$$\mathbf{P}\left\{T^{(d+2-2\gamma-2\varepsilon)/2}\left|\frac{\lambda(x,T)}{m^T} - p(0\rightsquigarrow x,T)B\right| \ge 1\right\} \le \exp(-O(T^\varepsilon)) \tag{4.29}$$

and

$$\lim_{T\to\infty} T^{d+2-2\gamma-\varepsilon}\mathbf{E}\left(\frac{\lambda(x,T)}{m^T} - p(0\rightsquigarrow x,T)B\right)^2 = 0. \tag{4.30}$$

Consequently for any fixed $x \in \mathbb{Z}^d$ and $0 < \varepsilon < 1$ we have

$$\lim_{T\to\infty} T^{1-\varepsilon}\left|\frac{\lambda(x,T)}{m^T p(0\rightsquigarrow x,T)} - B\right| = \lim_{T\to\infty} T^{1-\varepsilon}\left|\frac{1}{2}\left(\frac{2\pi T}{d}\right)^{d/2}\frac{\lambda(x,T)}{m^T} - B\right| = 0 \quad a.s. \tag{4.31}$$

provided that $x \equiv T \pmod 2$.

We also show that (4.31) gives the strongest possible rate in the following sense:

THEOREM 4.9 *For any $C > 0$ there exists a $\delta = \delta(C) > 0$ such that*

$$\mathbf{P}\left\{\left|\frac{1}{2}\left(\frac{4\pi T}{d}\right)^{d/2}\frac{\lambda(0,2T)}{m^{2T}} - B\right| \geq \frac{C}{T}\right\} \geq \delta. \tag{4.32}$$

Proof of Theorem 4.7. Let

$$\begin{aligned} T &= [t^\beta], \qquad \beta = 1/\varepsilon, \\ a(x,T) &= \frac{\lambda(x,T)}{m^T} - \frac{\mathbf{E}(\lambda(x,T) \mid \mathcal{F}(t))}{m^T}, \\ b(x,T) &= \frac{\mathbf{E}(\lambda(x,T) \mid \mathcal{F}(t))}{m^T} - p(0 \rightsquigarrow x,T)\frac{B(t)}{m^t}, \\ c(x,T) &= p(0 \rightsquigarrow x,T)\left(\frac{B(t)}{m^t} - B\right). \end{aligned}$$

Then by (4.21) and (4.23) with $\gamma \leq \alpha$ we have

$$\begin{aligned} &\mathbf{E}\left(\sum_{x\in D(aT^\gamma)}\left|\frac{\lambda(x,T)}{m^T} - p(0 \rightsquigarrow x,T)B\right|\right) \leq \\ &\leq \mathbf{E}\left(\sum_{x\in D(aT^\gamma)}|a(x,T)|\right) + \mathbf{E}\left(\sum_{x\in D(aT^\gamma)}|b(x,T)|\right) + \mathbf{E}\left(\sum_{x\in D(aT^\gamma)}|c(x,T)|\right) \leq \\ &\leq \exp(-O(T^\epsilon)) + \mathbf{E}\left(\sum_{x\in D(aT^\gamma)}|b(x,T)|\right). \end{aligned}$$

By Lemmas 4.9 and 4.10 we have

$$\begin{aligned} \sum_{x\in D(aT^\gamma)}|b(x,T)| &\leq \sum_{x\in D(aT^\gamma)} p(0 \rightsquigarrow x,T)O\left(\frac{|x|t+t^2}{T}\right) \leq \\ &\leq (aT^\gamma)^d O\left(\frac{|x|t+t^2}{T^{1+d/2}}\right) = O(T^{-(d+2-2\gamma(d+1)-\epsilon)/2}). \end{aligned}$$

Hence we have (4.27) and (4.28).

Proof of Theorem 4.8. Since by Lemmas 4.9 and 4.10

$$|b(x,T)| \leq p(0 \rightsquigarrow x,T)O\left(\frac{|x|t+t^2}{T}\right) \leq O\left(\frac{|x|t+t^2}{T^{1+d/2}}\right) = O(T^{-(d+2-2\gamma-\epsilon)/2})$$

we get (4.29) and (4.31) by Markov inequality repeating the proof of Theorem 4.7.

In order to prove (4.30) consider

$$\begin{aligned} &\mathbf{E}\left(\frac{\lambda(x,T)}{m^T} - p(0 \rightsquigarrow x,T)B\right)^2 = \mathbf{E}a^2(x,T) + \mathbf{E}b^2(x,T) + \mathbf{E}c^2(x,T) + \\ &\quad + 2\mathbf{E}a(x,T)b(x,T) + 2\mathbf{E}a(x,T)c(x,T) + 2\mathbf{E}b(x,T)c(x,T) \leq \\ &\leq \mathbf{E}a^2(x,T) + \mathbf{E}b^2(x,T) + \mathbf{E}c^2(x,T) + 2(\mathbf{E}a^2(x,T)\mathbf{E}b^2(x,T))^{1/2} + \\ &\quad + 2(\mathbf{E}a^2(x,T)\mathbf{E}c^2(x,T))^{1/2} + 2(\mathbf{E}b^2(x,T)\mathbf{E}c^2(x,T))^{1/2}. \end{aligned}$$

Since by Lemma 4.8

$$\mathbf{E}a^2(x,T) \leq C\frac{1}{m^t(T-t)^{d/2}},$$

by Lemmas 4.9 and 4.10

$$\mathbf{E}b^2(x,T) \leq (p(0 \rightsquigarrow x,T))^2 O\left(\left(\frac{xt+t^2}{T}\right)^2\right) \leq O\left(\frac{1}{T^d}\cdot\frac{T^{2\gamma+\epsilon}}{T^2}\right),$$

by (4.3)

$$\mathbf{E}c^2(x,T) \leq O(m^{-t})$$

we have (4.30).

Proof of Theorem 4.9 is based on the following observation.
Let $x = x(t,d) = (0,0,\ldots,0,2t) \in \mathbb{Z}^d$. Then for any fixed t

$$\mathbf{P}\{\lambda(x,2t) = B(2t)\} = p(t) > 0.$$

By Lemmas 4.3 and 4.10 we have

$$\mathbf{E}(\lambda(0,2T) \mid \lambda(x,2t) = B(2t)) = m^{2T-2t}p(x \rightsquigarrow 0, 2T-2t)B(2t) \sim$$
$$\sim m^{2T-2t}p(0 \rightsquigarrow 0,2T)B(2t)\left(1+O\left(\frac{t^2}{T}\right)\right). \tag{4.33}$$

Let

$$U = \frac{1}{2}\left(\frac{4\pi T}{d}\right)^{d/2}\frac{\lambda(0,2T)}{m^{2T}} - \frac{1}{2}\left(\frac{4\pi T}{d}\right)^{d/2}\frac{\mathbf{E}(\lambda(0,2T)\mid\lambda(x,2t)=B(2t))}{m^{2T}},$$
$$V = \frac{1}{2}\left(\frac{4\pi T}{d}\right)^{d/2}\frac{\mathbf{E}(\lambda(0,2T)\mid\lambda(x,2t)=B(2t))}{m^{2T}} - \frac{B(2t)}{m^{2t}},$$
$$W = \frac{B(2t)}{m^{2t}} - B.$$

Then

$$\frac{1}{2}\left(\frac{4\pi T}{d}\right)^{d/2}\frac{\lambda(0,2T)}{m^{2T}} - B = U+V+W$$

and by (4.33)

$$\mathbf{P}\left\{V \geq O\left(\frac{t^2}{T}\right) > 0\right\} \geq p(t).$$

Clearly there exist $c_1 > 0,\ c_2 > 0,\ c_3 > 0$ such that for any $0 < t < T < \infty$ we have

$$\mathbf{P}\{U > 0 \mid \lambda(x,2t) = B(2t)\} \geq c_1,$$
$$\mathbf{P}\{W > 0 \mid \lambda(x,2t) = B(2t)\} \geq c_2,$$
$$\mathbf{P}\left\{U>0,\ V \geq O\left(\frac{t^2}{T}\right),\ W>0 \,\middle|\, \lambda(x,2t)=B(2t)\right\} \geq c_3 p(t)$$

which easily implies (4.32).

(4.31) and Theorem 4.9 combined suggest the following

Conjecture.

$$\lim_{T\to\infty} \mathbf{P}\left\{T\left(\frac{1}{2}\left(\frac{4\pi T}{d}\right)^{d/2}\frac{\lambda(0,2T)}{m^{2T}} - B\right) < x\right\} = G(x)$$

with some distribution function $G(\cdot)$.

A further interesting question is to study the properties of the random set

$$A_T(d) = \{x:\ x \in \mathbb{Z}^d,\ \lambda(x,T) > 0\}.$$

Here we mention only a very weak theorem without proof. It can be proved by the methods of proofs of Theorems 11.3, 11.4 and 11.5.

THEOREM 4.10 *Let $d = 1$ and f_0 be the solution of the equation*

$$f_0 \log 2f_0 + (1 - f_0)\log(2(1 - f_0)) = \log m.$$

Then for any $f < f_0$

$$\lambda(2x, 2T) > 0 \qquad a.s.$$

for any $x \in (-fT, fT)$ and for all but finitely many T.

4.6 Independence and branching

Execute the following two experiments:

(i) Observe the values of the r.v.'s $\lambda(x,T)$ for any x and for a given large T.

(ii) Put $[Bm^T]$ particles at time $t = 0$ in $x = 0 \in \mathbb{Z}^d$. Let them move according independent random walks and observe the number of particles at time T in x. Denote this number by $\tilde{\lambda}(x,T)$.

Give these two numbers to a statistician and ask which one is due to experiment (i). Since

$$\mathbf{E}\lambda(x,T) = \mathbf{E}\tilde{\lambda}(x,T),$$

the question is non-trivial. However, it is easy to see that

$$|\tilde{\lambda}(0,2T) - m^{2T}Bp(0 \rightsquigarrow 0, 2T)| = O(m^T T^{-d/4})$$

and by Theorems 4.8 and 4.9

$$|\lambda(0,2T) - m^{2T}Bp(0 \rightsquigarrow 0, 2T)| = O(m^T T^{-(1+d/2)}).$$

Hence the statistician might say that: "I accept that number as the result produced by a branching random walk for which the number of particles in 0 is closer to $m^{2T}Bp(0 \rightsquigarrow 0, 2T)$."

4.7 Random lifetime

The model investigated in this Section is nearly the same as that of Section 4.2. The only difference is that in Section 4.2 the lifetime of each particle is 1 while here the lifetimes are governed by an array of i.i.d.r.v.'s.

We formulate this new model as follows. Let $E(x,t,\mu)$ ($x \in \mathbb{Z}^d$, $t = 0,1,2,\ldots$, $\mu = 1,2,\ldots,\lambda(x,t)$) be the lifetime of the μ-th particle located in x at time t according to the original model. Assume that $\{E(x,t,\mu)\}$ is an array of i.i.d.r.v's with

$$E(x,t,\mu) > 0, \quad \mathbf{E}E(x,t,\mu) = 1, \quad \mathbf{E}(E(x,t,\mu))^3 < \infty.$$

We also assume that the array $\{X(x,t,\mu), Z(x,t,\mu)\}$ of the original model is independent from the array $\{E(x,t,\mu)\}$.

We mean that the particle located in $x = 0$ at $t = 0$ lives up to time $E(0,0,1)$. At this time it jumps to $X(0,0,1)$, gives birth to $Z(0,0,1)$ children and dies. The μ-th particle ($\mu = 1,2,\ldots,Z(0,0,1) = \lambda(X(0,0,1),1)$ located in $X(0,0,1)$ at $E(0,0,1)$ lives up to $E(X(0,0,1),1,\mu)$, moves to $X(X(0,0,1),1,\mu)$, produces $Z(X(0,0,1),1,\mu)$ children and dies. Continuing this procedure we get a new process. It is called *branching random walk with random lifetime*. Let $\bar{\lambda}(x,t)$ be the number of particles located in x at time t in this new process. Note that $\bar{\lambda}(0,t) = 1$ if $t < E(0,0,1)$, $\bar{\lambda}(x,t) = Z(0,0,1)$ if $x = X(0,0,1)$ and

$$E(0,0,1) < t < E(0,0,1) + \min_{\mu} E(x,1,\mu)$$

and so on. Since $\mathbf{E}E(x,t,\mu) = 1$ we might expect that the limit behaviour of $\bar{\lambda}(x,t)$ is the same as that of $\lambda(x,t)$. It turns out that it is not true at all. In fact we have

THEOREM 4.11 *For any $\alpha > 0$*

$$\limsup_{t\to\infty} \frac{\bar{\lambda}(0,t)}{m^{t+\alpha t^{1/2}}} = \infty \qquad a.s.$$

Proof. Consider the

$$\lambda\left(0, 2\left[\frac{t+\alpha t^{1/2}}{2}\right]\right)$$

in the original model. Then by Theorem 4.8

$$\lambda(0,T) \sim 2Bm^T\left(\frac{d}{2\pi T}\right)^{d/2} \sim O(m^{t+\alpha t^{1/2}} t^{-d/2})$$

where

$$T = 2\left[\frac{t+\alpha t^{1/2}}{2}\right].$$

Consider the $\lambda(0,T)$ particles located in 0 at time T in the original model. In the new model the time needed for T steps is the sum of T i.i.d.r.v.'s. Hence the probability, that in the new model the required time is between $t-1$ and t, is $O(t^{-1/2})$. Consequently with positive probability

$$\bar{\lambda}(0,t) \geq O(m^{t+\alpha t^{1/2}} t^{-(d/2+1)})$$

which implies Theorem 4.11.

4.8 Generalizations

Let π be a positive integer valued random variable with

$$\mathbf{P}\{\pi = k\} = q_k \qquad (k = 1, 2, \ldots)$$

where

$$q_k \geq 0, \qquad \sum_{k=1}^{\infty} q_k = 1,$$

$$\sum_{k=1}^{\infty} k q_k = \mu, \qquad \sum_{k=1}^{\infty} (k-\mu)^2 q_k = \Theta^2 < \infty.$$

Then we might generalize our original model (Section 4.2) as follows. At time $t = 0$ there exist π particles located in $0 \in \mathbb{Z}^d$ Assume that these π particles perform independent branching random walks with a branching distribution $\{p_k\}$ satisfying the conditions of Section 4.1. Let $\lambda^*(x,t)$ $(x \in \mathbb{Z}^d,\ t = 0, 1, 2, \ldots)$ be the number of particles in x at t. Clearly

$$\lambda^*(x,0) = \begin{cases} \pi & \text{if} \quad x = 0, \\ 0 & \text{if} \quad x \neq 0. \end{cases}$$

Let $B_i(t)$ $(i = 1, 2, \ldots, \pi;\ t = 0, 1, 2, \ldots)$ be the number of offsprings of the i-th particle after t steps and let

$$\lim_{t \to \infty} \frac{B_i(t)}{m^t} = B_i$$

where

$$\mathbf{P}\{B_i(1) = k\} = p_k, \quad (k = 0, 1, 2, \ldots;\ i = 1, 2, \ldots), \qquad m = \sum_{k=0}^{\infty} k p_k$$

and $B_1, B_2, \ldots$ are i.i.d.r.v.'s. Let

$$B_1 + B_2 + \cdots + B_\pi = B^*.$$

Then as a simple consequence of the above results we have

THEOREM 4.12 *Replacing $\lambda(\cdot,\cdot)$ by $\lambda^*(\cdot,\cdot)$ and B by B^* Theorems 4.4 – 4.8 remain true.*

This model can be further generalized as follows. Assume that the π particles, located in $x = 0$ at $t = 0$, have different kind of fertilities. We mean the following: let

$$p_i = (p_{i0}, p_{i1}, \ldots) \qquad (i = 1, 2, \ldots)$$

with

$$p_{ij} \geq 0, \qquad \sum_{j=0}^{\infty} p_{ij} = 1,$$

$$\sum_{j=0}^{\infty} j p_{ij} = m_i > 1, \qquad \sum_{j=0}^{\infty} (j - m_i)^2 p_{ij} = \sigma_i^2 < \infty$$

and p_i characterizes the fertility of the i-th particle i.e. if $B_i(1)$ is the number of the offsprings of the i-th particle then

$$\mathbf{P}\{B_i(1) = j\} = p_{ij} \qquad (i = 1, 2, \ldots, \pi;\ j = 0, 1, 2, \ldots).$$

We assume that the fertilities of the offsprings of the i-th particle are also generated by the distribution p_i. Let

$$\begin{aligned} \hat{m} &= \max(m_1, m_2, \ldots, m_\pi), \\ \mathcal{A} &= \{i :\ 1 \leq i \leq \pi,\ m_i = \hat{m}\}, \\ K &= |\mathcal{A}|, \end{aligned}$$

$B_i(t)$ be the number of offsprings of the i-th particle at t $(t = 0, 1, 2, \ldots;$ $B_i(0) = 1)$

$$\lim_{t\to\infty} \frac{B_i(t)}{m_i^t} = B_i,$$

$\tilde{\lambda}(x, t)$ $(x \in \mathbb{Z}^d,\ t = 0, 1, 2, \ldots)$ be the number of particles in x at t,

$$\tilde{B} = \sum_{i\in\mathcal{A}} B_i.$$

Clearly

$$\tilde{\lambda}(x, 0) = \begin{cases} \pi & \text{if} \quad x = 0, \\ 0 & \text{if} \quad x \neq 0, \end{cases}$$

$$\sum_{x\in\mathbb{Z}^d} \tilde{\lambda}(x, t) = \sum_{i=1}^{\pi} B_i(t).$$

Then as a simple consequence of the above results we have

THEOREM 4.13 *Replacing $\lambda(\cdot,\cdot)$ by $\tilde{\lambda}(\cdot,\cdot)$ and B by $\tilde{B}$ Theorems 4.4 – 4.8 remain true.*

Chapter 5

Branching random walks of a random field

5.1 The model

At time $t = 0$ one particle is located in each $x \in \mathbb{Z}^d$. At time $t = 1$ each of these particles move independently to one of their $2d$ neighbours with probability $1/2d$ and produces i ($i = 0, 1, 2, \ldots$) offsprings with probability p_i and dies. This procedure will be repeated at each time unit.

In other words at $t = 0$ from each $x \in \mathbb{Z}^d$ a branching random walk is started and these branching random walks are independent. Let $\lambda_x(y,t)$ ($x \in \mathbb{Z}^d$; $y \in \mathbb{Z}^d$; $t = 0, 1, 2, \ldots$) be the number of those offsprings of the particle located in x at time $t = 0$ which are located in y at time t. Note that for any $x \in \mathbb{Z}^d$

$$\{\lambda_x(y,t) : y \in \mathbb{Z}^d;\ t = 0, 1, 2, \ldots\} \stackrel{\mathcal{D}}{=} \{\lambda(y - x, t) : y \in \mathbb{Z}^d;\ t = 0, 1, 2, \ldots\}$$

where $\lambda(\cdot, \cdot)$ is the branching random walk considered in Chapter 4.

Let

$$\Lambda(y,t) = \sum_{x \in \mathbb{Z}^d} \lambda_x(y,t), \qquad \Lambda(t) = \Lambda(0,t),$$

$$B_x(t) = \sum_{y \in \mathbb{Z}^d} \lambda_x(y,t)$$

and

$$\mathcal{F}_x(T) = \mathcal{F}\{\lambda_x(y,t);\ y \in \mathbb{Z}^d,\ t = 0, 1, 2, \ldots, T\} \qquad (x \in \mathbb{Z}^d)$$

be the smallest σ-algebra with respect to which the array in brackets is measurable.

Clearly

(i) $\Lambda(y,t)$ is the total number of particles located in $y \in \mathbb{Z}^d$ at time t,

(ii) $\Lambda(t)$ is the total number of particles located in $0 \in \mathbb{Z}^d$ at time t,

(iii) $B_x(t)$ is the total number of offsprings of the particle located in x at time $t = 0$, after t steps.

Note that

(i)

$$\Lambda(y,0) = \Lambda(0) = 1 \text{ for any } y \in \mathbb{Z}^d,$$

(ii)
$$\lambda_x(y,0) = \begin{cases} 1 & \text{if} \quad x = y, \\ 0 & \text{if} \quad x \neq y, \end{cases}$$

(iii) for any $x \in \mathbb{Z}^d$ $\{B_x(t) : t = 0,1,2,\ldots\}$ is a branching process and the processes $\{B_{x_i}(t) : t = 0,1,2,\ldots\}$ $(i = 1,2,\ldots,n;\ n = 2,3,\ldots)$ are independent whenever $x_i \neq x_j$ $(i \neq j)$.

From now on it will be assumed again that

$$p_i \geq 0, \qquad \sum_{i=0}^{\infty} p_i = 1, \qquad \sum_{i=0}^{\infty} ip_i = m > 1,$$

$$\sum_{i=0}^{\infty}(i-m)^2 p_i = \sigma^2 < \infty.$$

For sake of simplicity in this Chapter we also assume that

$$\sum_{i=0}^{\infty}(i-m)^4 p_i < \infty.$$

By Theorem 4.2 for any $x \in \mathbb{Z}^d$ there exists a nonnegative r.v. B_x such that

$$\lim_{t\to\infty} \frac{B_x(t)}{m^t} = B_x \qquad \text{a.s.}$$

and the r.v.'s B_x $(x \in \mathbb{Z}^d)$ are independent.

In this Chapter we intend to investigate the properties of the process $\Lambda(y,t)$ as $t \to \infty$.

5.2 A LIL and a CLT

At first we present a strong law of large numbers.

THEOREM 5.1

$$\lim_{T\to\infty} \frac{\Lambda(T)}{m^T} = 1 \qquad a.s.$$

In fact we prove the following much stronger law of iterated logarithm.

THEOREM 5.2

$$\limsup_{T\to\infty} \frac{T^{d/4}}{\vartheta(2\log\log T)^{1/2}} \left|\frac{\Lambda(T)}{m^T} - 1\right| = 1 \qquad a.s$$

where

$$\vartheta = 2^{1/2}\delta\left(\frac{d}{4\pi}\right)^{d/4}$$

and

$$\delta = \frac{\sigma}{(m^2-m)^{1/2}}.$$

Proof. Introduce the following notations:

$$\Delta(x,T) = \frac{\lambda_x(0,T)}{m^T} - p(x \rightsquigarrow 0,T)B_x,$$
$$a_x(T) = \frac{\lambda_x(0,T)}{m^T} - \frac{\mathbf{E}(\lambda_x(0,T) \mid \mathcal{F}_x(t))}{m^T},$$
$$b_x(T) = \frac{\mathbf{E}(\lambda_x(0,T) \mid \mathcal{F}_x(t))}{m^T} - p(x \rightsquigarrow 0,T)\frac{B_x(t)}{m^t},$$
$$c_x(T) = p(x \rightsquigarrow 0,T)\left(\frac{B_x(t)}{m^t} - B_x\right).$$

where $p(x \rightsquigarrow 0,T) = p(0 \rightsquigarrow x,T)$ is defined by (ii) of Section 4.3 and $t = T^\epsilon$.

Let $1/2 < \gamma < 1$. Then we have

$$\frac{\Lambda(T)}{m^T} = \frac{1}{m^T}\sum_{x\in\mathbb{Z}^d}\lambda_x(0,t) =$$
$$= \frac{1}{m^T}\sum_{x\in C(0,T)}\lambda_x(0,T) = \sum_{x\in C(0,T)} p(x \rightsquigarrow 0,T)B_x + \sum_{x\in C(0,T)}\Delta(x,T) =$$
$$= \sum_{x\in C(0,T)} p(x \rightsquigarrow 0,T)B_x + \sum_{x\in C(0,T^\gamma)}\Delta(x,T) + \sum_{x\in C(0,T)-C(0,T^\gamma)}\Delta(x,T). \quad (5.1)$$

Observe that by (4.2) of Theorem 4.2, (4.6) of Lemma 4.5 and (4.19) we have

$$\sum_{x\in C(0,T)-C(0,T^\gamma)} \mathbf{E}|\Delta(x,T)| \le$$
$$\le \frac{1}{m^T}\sum_{x\in C(0,T)-C(0,T^\gamma)}\mathbf{E}\lambda_x(0,T) + \sum_{x\in C(0,T)-C(0,T^\gamma)} p(x \rightsquigarrow 0,T) =$$
$$= 2\sum_{x\in C(0,T)-C(0,T^\gamma)} p(x \rightsquigarrow 0,T) \le \exp(-O(T^{2\gamma-1})). \quad (5.2)$$

Taking into account that the processes $a_x(\cdot)$ $(x \in \mathbb{Z}^d)$ are independent, by Lemma 4.8 we have

$$\mathbf{E}\left(T^{d/4}\sum_{x\in C(0,T^\gamma)} a_x(T)\right)^2 \le T^{d/2}\frac{CT^{\gamma d}}{m^t(T-t)^{d/2}}. \quad (5.3)$$

By Lemmas 4.9 and 4.10

$$|b_x(T)| \le p(x \rightsquigarrow 0,T)\frac{B_x(t)}{m^t}O\left(\frac{|x|t+t^2}{T}\right) \le \frac{B_x(t)}{m^t}O\left(\frac{|x|t+t^2}{T^{1+d/2}}\right)$$

i.e.

$$\mathbf{E}|b_x(T)| \le O\left(\frac{|x|t+t^2}{T^{1+d/2}}\right),$$
$$\mathbf{E}b_x^2(T) \le O\left(\frac{(|x|t+t^2)^2}{T^{2+d}}\right)$$

and

$$\mathbf{E}b_x^4(T) \le \left(\frac{(|x|t+t^2)^4}{T^{4+2d}}\right).$$

Taking into account that the processes $b_x(\cdot)$ are independent and choosing $\gamma = 1/2+\varepsilon$ and $\varepsilon > 0$ small enough we have

$$T^d\mathbf{E}\left(\sum_{x\in C(0,T^\gamma)} b_x(T)\right)^4 \le O(T^{-1-\varepsilon}). \tag{5.4}$$

By (4.3)

$$\mathbf{E}\left(T^{d/4}\sum_{x\in C(0,T^\gamma)} c_x(T)\right)^2 \le \exp(-O(T^\varepsilon)). \tag{5.5}$$

(5.2), (5.3), (5.4) and (5.5) combined imply

$$\lim_{T\to\infty} T^{d/4}\sum_{x\in C(0,T)} \Delta(x,T) = 0 \qquad \text{a.s.}$$

Observe that

$$\mathbf{E}\left(\sum_{x\in C(0,T)} p(x\rightsquigarrow 0,T)B_x\right) = 1$$

and

$$\operatorname{Var}\left(\sum_{x\in C(0,T)} p(x\rightsquigarrow 0,T)B_x\right) \sim 2\delta^2\left(\frac{d}{4\pi T}\right)^{d/2}.$$

Hence in order to prove Theorem 5.2 it is enough to prove that

$$\limsup_{T\to\infty}\frac{T^{d/4}}{\vartheta(2\log\log T)^{1/2}}\left|\sum_{x\in C(0,T)} p(x\rightsquigarrow 0,T)B_x - 1\right| = 1 \qquad \text{a.s.}$$

This follows from the LIL for weighted sums (cf. [52]).

Note that Theorem 5.2 implies that the rate of convergence in Theorem 5.1 is better if d is larger.

THEOREM 5.3

$$\lim_{T\to\infty}\mathbf{P}\left\{\frac{T^{d/4}}{\vartheta}\left(\frac{\Lambda(T)}{m^T}-1\right) < x\right\} = \Phi(x).$$

Proof is the same as that of Theorem 5.2 applying the corresponding CLT.

5.3 On the locations of nonoccupied points

Theorem 5.1 clearly implies that for any $x \in \mathbb{Z}^d$

$$\lim_{T\to\infty} \Lambda(x,T) = \infty \qquad \text{a.s.}$$

However, it is also clear that for any $T = 1, 2, \ldots$ we have

$$\inf_{x\in\mathbb{Z}^d} \Lambda(x,T) = 0 \qquad \text{a.s.}$$

It looks interesting to study the properties of the set of those x's for which $\Lambda(x,T) = 0$. Here we consider only the case $d = 1$ and prove

THEOREM 5.4 *Let*

$$M(T) = \min\{x : \ x \in \mathbb{Z}^1, \ x \geq 0, \ \Lambda(x,T) = 0\}.$$

Then for any $\varepsilon > 0$ and for all but finitely many T we have

$$\exp\left((1-\varepsilon)\left(\log\frac{1}{q}\right) f_0 T\right) \leq M(T) \leq \exp\left(2(1+\varepsilon)\left(\log\frac{1}{q}\right) T\right).$$

where q resp. f_0 are defined in Theorem 4.1 resp. 4.10.

Proof. At first we prove that for any $\varepsilon > 0$

$$\frac{\log M(T)}{T} \leq 2(1+\varepsilon)\log\frac{1}{q} \quad \text{a.s.} \tag{5.6}$$

for all but finitely many T. The proof of (5.6) is based on the following:

LEMMA 5.1 ([44] Theorem 7.1) *Let $X_1, X_2, \ldots$ be a sequence of i.i.d.r.v.'s with*

$$\mathbf{P}\{X_1 = 0\} = 1 - \mathbf{P}\{X_1 = 1\} = p \qquad (0 < p < 1).$$

Further let

$$S_0 = 0, \qquad S_n = X_1 + X_2 + \cdots + X_n, \qquad (n = 1, 2, \ldots)$$

$$K_N = -\frac{\log N}{\log p}.$$

Then for any $0 < \varepsilon < 1$

$$\min_{0\leq n\leq N-K}(S_{n+K} - S_n) = 0 \quad \textit{if} \quad K \leq (1-\varepsilon)K_N \qquad \textit{a.s.}$$

and

$$\min_{0\leq n\leq N-K}(S_{n+K} - S_n) \geq 1 \quad \textit{if} \quad K \geq (1+\varepsilon)K_N \qquad \textit{a.s.}$$

for all but finitely many N.

In [44] only the case $p = 1/2$ is treated. However the general case can be settled on the same way.

By (4.1) for any $\varepsilon > 0$ there exists a positive integer $\nu = \nu(\varepsilon)$ such that

$$\mathbf{P}\{B_x(t) = 0,\ t \geq \nu\} \geq q - \varepsilon \qquad (x \in \mathbb{Z}^1).$$

Hence by Lemma 5.1 there exists an $x_0 \in \mathbb{Z}^1$ such that

$$T \leq x_0 \leq \exp\left(2(1+\varepsilon)T \log \frac{1}{q-\varepsilon}\right) - T$$

and

$$B_x(t) = 0$$

for any $t \geq \nu$ and $|x_0 - x| \leq T$. This clearly implies (5.6).

Now we prove that for any $\varepsilon > 0$

$$\frac{\log M(T)}{T} \geq (1-\varepsilon)\left(\log \frac{1}{q}\right) f_0 \qquad \text{a.s.} \tag{5.7}$$

for all but finitely many T.

Let

$$2x \in [0, \exp(2cT)] \qquad (c > 0,\ x \in \mathbb{Z}^1).$$

Then by Lemma 5.1 there exists a $j = j(x, \omega)$ such that

$$2j \in (2x - dT, 2x + dT) \qquad (d > 0,\ j \in \mathbb{Z}^1)$$

and

$$B_{2j} > 0$$

provided that

$$d < \frac{c}{\log 1/q}.$$

By Theorem 4.10

$$\Lambda(2x, 2T) \geq \lambda_{2j}(2x, 2T) > 0$$

if $d > f_0$. Hence we have (5.7) and Theorem 5.4 is proved.

5.4 The covariance of $\Lambda(x, T)$

THEOREM 5.5 *For any $x \in \mathbb{Z}^d$, $y \in \mathbb{Z}^d$ we have*

$$\mathbf{E}((m^{-T}\Lambda(x,T) - 1)(m^{-T}\Lambda(y,T) - 1)) = (\delta^2 + o(1))p(x \rightsquigarrow y, 2T)$$

where $o(1)$ does not depend on x and y.

Proof. Repeating the proof of Theorem 5.2 we get

$$\mathbf{E}((m^{-T}\Lambda(x,T)-1)(m^{-T}\Lambda(y,T)-1)) \sim$$
$$\sim \mathbf{E}\left(\sum_{u\in\mathbb{Z}^d} p(u \rightsquigarrow x,T)(B_u-1)\right)\left(\sum_{v\in\mathbb{Z}^d} p(v \rightsquigarrow y,T)(B_v-1)\right) =$$
$$= \delta^2 \sum_{u\in\mathbb{Z}^d} p(u \rightsquigarrow x,T)p(u \rightsquigarrow y,T) = \delta^2 p(x \rightsquigarrow y, 2T).$$

Hence we have Theorem 5.5.

Making use the method of proof of Theorem 5.2 it is easy to see that the joint limit distribution of

$$\left\{\frac{T^{d/4}}{\vartheta}\left(\frac{\Lambda(x_i,T)}{m^T}-1\right),\ i=1,2,\ldots,n\right\} \qquad (n=1,2,\ldots)$$

is Gaussian with the covariance given above. This easily implies that the processes

$$\left\{\frac{T^{d/4}}{\vartheta}\left(\frac{\Lambda(xT^{1/2},T)}{m^T}-1\right),\ x\in\mathbb{Z}^d\right\}$$

weakly converges (as $T\to\infty$) to a Gaussian process $\{\Gamma(x),\ x\in\mathbb{Z}^d\}$ with

$$\mathbf{E}\Gamma(x)=0, \qquad \mathbf{E}\Gamma(x)\Gamma(y)=\exp\left(-\frac{\|x-y\|^2}{2}\right).$$

5.5 A generalization

Let $\{\pi_x,\ x\in\mathbb{Z}^d\}$ be an array of nonnegative integer valued i.i.d.r.v.'s with

$$\mathbf{P}\{\pi_x=k\}=q_k \qquad (k=0,1,2,\ldots,\ x\in\mathbb{Z}^d),$$
$$q_k\geq 0, \qquad \sum_{k=0}^{\infty} q_k = 1,$$
$$\sum_{k=0}^{\infty} kq_k=\mu, \qquad \sum_{k=0}^{\infty}(k-\mu)^2 q_k=\Theta^2<\infty.$$

Then we might generalize our original model (Section 5.1) as follows. At time $t=0$ π_x particles located in $x\in\mathbb{Z}^d$. At time $t=1$ each of these particles move independently to one of their $2d$ neighbours with probability $1/2d$ and produce i $(i=0,1,2,\ldots)$ offsprings with probability p_i and die. Assume that the offspring-distribution $\{p_i\}$ satisfies the conditions of Section 5.1. This procedure will be repeated at each time unit.

In other words at time $t=0$ from $x\subset\mathbb{Z}^d$ π_x branching random walks are started independently. Let

$$\lambda_i(x \rightsquigarrow y,t) \qquad (x\in\mathbb{Z}^d,\ y\in\mathbb{Z}^d,\ i=1,2,\ldots,\pi_x,\ t=0,1,2,\ldots)$$

be the number of those offsprings of the i-th particle located in x at time $t = 0$ which are located in y at time t. Note that for any $x \in \mathbb{Z}^d$ and $i = 1, 2, \ldots, \pi_x$ we have

$$\{\lambda_i(x \rightsquigarrow y, t) : \ y \in \mathbb{Z}^d, \ t = 0, 1, 2, \ldots\} \stackrel{\mathcal{D}}{=} \{\lambda(y - x, t) : \ y \in \mathbb{Z}^d, \ t = 0, 1, 2, \ldots\}$$

where $\lambda(\cdot, \cdot)$ is the branching random walk considered in Chapter 4.

Let

$$\begin{aligned}
\lambda^*(x \rightsquigarrow y, t) &= \sum_{i=1}^{\pi_x} \lambda_i(x \rightsquigarrow y, t), \\
\Lambda^*(y, t) &= \sum_{x \in \mathbb{Z}^d} \lambda^*(x \rightsquigarrow y, t), \\
\Lambda^*(t) &= \Lambda(0, t), \\
B_i(x, t) &= \sum_{y \in \mathbb{Z}^d} \lambda_i(x \rightsquigarrow y, t), \\
\sum_{i=1}^{\pi_x} B_i(x, t) &= B^*(x, t).
\end{aligned}$$

Clearly

(i) $\lambda^*(x \rightsquigarrow y, t)$ is the number of those offsprings of the particles located in x at $t = 0$ which are located in y at time t,

(ii) $\Lambda^*(y, t)$ is the total number of particles located in y a time t,

(iii) $B_i(x, t)$ is total number of offsprings of the i-th particle located in x at $t = 0$, after t steps,

(iv) $B^*(x, t)$ is total number of offsprings of the particles located in x at $t = 0$, after t steps.

Repeating the proofs of Theorems 5.2 and 5.3 we get

THEOREM 5.6 *Replacing $\Lambda(\cdot)$ by $\Lambda^*(\cdot)$ and ϑ by*

$$\Delta = \mu\vartheta^2 + m^2\Theta^2$$

Theorems 5.2 and 5.3 remain true.

Chapter 6

Branching Wiener process starting with one particle

6.1 The model

Here we modify the model of Section 4.2 replacing the simple, symmetric random walk by a Wiener process. In other words we have the following scheme:

(i) a particle starts from the position $0 \in \mathbb{R}^d$ and executes a Wiener process $W(t) \in \mathbb{R}^d$,

(ii) arriving at $t = 1$ to the new location $W(1)$, it dies,

(iii) at death it is replaced by Y offsprings where

$$\mathbf{P}\{Y = \ell\} = p_\ell \qquad (\ell = 0, 1, 2, \ldots)$$

and the offspring-distribution $\{p_\ell\}$ satisfies the conditions of Section 4.2,

(iv) each offspring, starting from where its ancestor dies, executes a Wiener process (from its starting point) and repeats the above given steps and so on. All Wiener processes and offspring-numbers are assumed independent of one-another.

In order to give a more formal definition, let

$$\{N(x,t,j), Z(x,t,j);\ x \in \mathbb{R}^d,\ t = 0,1,2,\ldots,\ j = 1,2,\ldots\}$$

be a set of independent r.v.'s with

$$\mathbf{P}\{N(x,t,j) < u\} = \Phi(u) \qquad (x \in \mathbb{R}^d,\ u \in \mathbb{R}^d),$$
$$\mathbf{P}\{Z(x,t,j) = k\} = p_k \qquad (k = 0,1,2,\ldots)$$

where

$$p_k \geq 0, \qquad \sum_{k=0}^{\infty} p_k = 1, \qquad \sum_{k=0}^{\infty} k p_k = m > 1,$$

$$0 < \sum_{k=0}^{\infty} (k-m)^2 p_k = \sigma^2 < \infty,$$

$$I_A(u) = \begin{cases} 1 & \text{if} \quad u \in A, \\ 0 & \text{if} \quad u \notin A, \end{cases}$$

$$I_x(u) = I_{\{x\}}(u) = \begin{cases} 1 & \text{if} \quad u = x, \\ 0 & \text{if} \quad u \neq x, \end{cases}$$

$$\lambda(x,0) = I_0(x).$$

Further for any Borel set $A \subset \mathbb{R}^d$ let

$$\psi(A,0) = I_A(0).$$

By induction we define $\lambda(x,t)$ and $\psi(A,t)$ as follows

$$\lambda(x,t) = \sum_{y \in \mathbb{R}^d} \sum_{j=1}^{\lambda(y,t-1)} I_x(N(y,t-1,j)+y)Z(y,t-1,j),$$

$$\psi(A,t) = \sum_{y \in \mathbb{R}^d} \sum_{j=1}^{\lambda(y,t-1)} I_A(N(y,t-1,j)+y)Z(y,t-1,j) = \sum_{x \in A} \lambda(x,t).$$

Clearly $\lambda(x,t)$ is the number of particles located in x at time t. Hence $\lambda(x,t) = 0$ for all but finitely many $x \in \mathbb{R}^d$, $t = 0,1,2,\ldots$. $\psi(A,t)$ is the number of particles located in A at time t.

6.2 On the moments of $\psi(A,t)$

Introduce the following notations:

$$H(A,y,t) = \begin{cases} (2\pi t)^{-d/2} \int_A \exp\left(-\frac{\|x-y\|^2}{2t}\right) dx & \text{if} \quad t > 0, \\ 1 & \text{if} \quad t = 0 \text{ and } y \in A, \\ 0 & \text{if} \quad t = 0 \text{ and } y \notin A, \end{cases}$$

$$F(A,T,t) = m^{T-t} \int_{\mathbb{R}^d} H(A,y,T-t)\psi(dy,t) = m^{T-t} \sum_{y \in \mathbb{R}^d} H(A,y,T-t)\lambda(y,t).$$

and

$$\mathcal{F}(T) = \mathcal{F}\{\lambda(x,t): \ x \in \mathbb{R}^d, \ t = 0,1,2,\ldots,T\} = \\ = \mathcal{F}\{\psi(A,t): \ A \subset \mathbb{R}^d, \ t = 0,1,2,\ldots,T\}$$

is the smallest σ-algebra with respect to which the r.v.'s $\lambda(x,t)$ ($x \in \mathbb{R}^d$, $t = 0,1,2,\ldots,T$) are measurable.

LEMMA 6.1 *For any* $t = 1,2,\ldots$ *we have*

$$\mathbf{E}(\psi(A,t) \mid \mathcal{F}(t-1)) = F(A,t,t-1) \tag{6.1}$$

and

$$\mathbf{E}\psi(A,t) = m^t H(A,0,t). \tag{6.2}$$

Proof. In case $t = 1$ by the definition of ψ we have

$$\psi(A,1) = I_A(N(0,0,1))Z(0,0,1)$$

and

$$\mathbf{E}(\psi(A,1) \mid \mathcal{F}(0)) = \mathbf{E}\psi(A,1) = mH(A,0,1) = F(A,1,0) \tag{6.3}$$

i.e. (6.1) and (6.2) for $t = 1$ are proved. Similarly

$$\begin{aligned}
&\mathbf{E}(\psi(A,2) \mid \lambda(\cdot,1)) = \mathbf{E}(\psi(A,2) \mid \mathcal{F}(1)) = \\
&= \mathbf{E}\left(\sum_{y\in\mathbb{R}^d}\sum_{j=1}^{\lambda(y,1)} I_A(N(y,1,j)+y)Z(y,1,j) \mid \lambda(\cdot,1)\right) = \\
&= m\sum_{y\in\mathbb{R}^d} H(A,y,1)\lambda(y,1) = m\int_{\mathbb{R}^d} H(A,y,1)\psi(dy,1) = F(A,2,1).
\end{aligned}$$

Hence by (6.3)

$$\mathbf{E}\psi(A,2) = m^2\int_{\mathbb{R}^d} H(A,y,1)H(dy,0,1) = m^2H(A,0,2).$$

Now, we prove our Lemma by induction. Assume that

$$\mathbf{E}\psi(A,t-1) = m^{t-1}H(A,0,t-1) \tag{6.4}$$

and consider

$$\begin{aligned}
&\mathbf{E}(\psi(A,t) \mid \lambda(\cdot,t-1)) = \\
&= \mathbf{E}\left(\sum_{y\in\mathbb{R}^d}\sum_{j=1}^{\lambda(y,t-1)} I_A(N(y,t-1,j)+y)Z(y,t-1,j) \mid \lambda(\cdot,t-1)\right) = \\
&= m\sum_{y\in\mathbb{R}^d} H(A,y,1)\lambda(y,t-1) = m\int_{\mathbb{R}^d} H(A,y,1)\psi(dy,t-1) = \\
&= F(A,t,t-1).
\end{aligned} \tag{6.5}$$

Hence we have (6.1). Since by (6.4) and (6.5)

$$\begin{aligned}
\mathbf{E}\psi(A,t) &= m\int_{\mathbb{R}^d} H(A,y,1)\mathbf{E}(\psi(dy,t-1)) = \\
&= m^t\int_{\mathbb{R}^d} H(A,y,1)H(dy,0,t-1) = m^tH(A,0,t),
\end{aligned}$$

we have (6.2).

LEMMA 6.2 *For any $\varepsilon > 0$, $t > 0$ and $\xi \in \mathbb{R}^d$ we have*

$$\frac{m^t}{(2\pi t)^{d/2}}|C(0,\varepsilon t)|\exp\left(-\frac{(\|\xi\|+\varepsilon)^2}{2}t\right) \le \mathbf{E}\psi(C(\xi t,\varepsilon t),t) \le \frac{m^t}{(2\pi t)^{d/2}}|C(0,\varepsilon t)|.$$

Assuming that $\|\xi\| > \varepsilon$ we also have

$$\mathbf{E}\psi(C(\xi t,\varepsilon t),t) \le \frac{m^t}{(2\pi t)^{d/2}}|C(0,\varepsilon t)|\exp\left(-\frac{(\|\xi\|-\varepsilon)^2}{2}t\right).$$

Proof. It is a trivial consequence of (6.2).

LEMMA 6.3

$$\mathbf{E}(\psi(A,T) \mid \mathcal{F}(t)) = F(A,T,t) \qquad (0 \le t < T).$$

Proof. By (6.1) we have

$$\begin{aligned}
&\mathbf{E}(F(A,T,t) \mid \mathcal{F}(t-1)) = \\
&= m^{T-t} \int_{\mathbb{R}^d} H(A,y,T-t)\mathbf{E}(\psi(dy,t) \mid \mathcal{F}(t-1)) = \\
&= m^{T-t} \int_{\mathbb{R}^d} H(A,y,T-t)F(dy,t,t-1) = \\
&= m^{T-t} \int_{\mathbb{R}^d} H(A,y,T-t)m \int_{\mathbb{R}^d} H(dy,u,1)\psi(du,t-1) = \\
&= m^{T-t+1} \int_{\mathbb{R}^d}\int_{\mathbb{R}^d} H(A,y,T-t)H(dy,u,1)\psi(du,t-1) = \\
&= m^{T-t+1} \int_{\mathbb{R}^d} H(A,u,T-t+1)\psi(du,t-1) = F(A,T,t-1).
\end{aligned}$$

Hence applying (6.1) again

$$\begin{aligned}
&\mathbf{E}(\psi(A,T) \mid \mathcal{F}(t)) = \\
&= \mathbf{E}(\mathbf{E}(\psi(A,T) \mid \mathcal{F}(T-1)) \mid \mathcal{F}(t)) = \mathbf{E}(F(A,T,T-1) \mid \mathcal{F}(t)) = \\
&= \mathbf{E}(\mathbf{E}(F(A,T,T-1) \mid \mathcal{F}(T-2)) \mid \mathcal{F}(t)) = \mathbf{E}(F(A,T,T-2) \mid \mathcal{F}(t)) = \cdots = \\
&= \mathbf{E}(F(A,T,t) \mid \mathcal{F}(t)) = F(A,T,t).
\end{aligned}$$

Observe that Lemmas 6.3 is the natural analogue of (4.4).

Now, we formulate the analogues of the other lemmas of Section 4.3.

LEMMA 6.4 (cf. (4.7))

$$\mathbf{E}(F(A,T,t) \mid \mathcal{F}(t-1)) = F(A,T,t-1).$$

Proof. By Lemma 6.3

$$\begin{aligned}
\mathbf{E}(F(A,T,t) \mid \mathcal{F}(t-1)) &= \mathbf{E}(\mathbf{E}(\psi(A,T) \mid \mathcal{F}(t)) \mid \mathcal{F}(t-1)) = \\
&= \mathbf{E}(\psi(A,T) \mid \mathcal{F}(t-1)) = F(A,T,t-1)
\end{aligned}$$

and we have Lemma 6.4.

LEMMA 6.5 *Let $A_1, A_2, \dots$ be a partition of $\mathbb{R}^d$. Then*

$$\begin{aligned}
&\sum_{k=1}^{\infty} \operatorname{Var}(\psi(A_k,T) \mid \mathcal{F}(T-1)) = \\
&= \sum_{k=1}^{\infty} \mathbf{E}((\psi(A_k,T) - \mathbf{E}(\psi(A_k,T) \mid \mathcal{F}(T-1)))^2 \mid \mathcal{F}(T-1)) \le \\
&\le B(T-1)(m^2+\sigma^2)
\end{aligned}$$

and for any Borel set $A \subset \mathbb{R}^d$ and $k = 1, 2, \ldots$ we have

$$\mathrm{Var}(\psi(A,T) \mid \mathcal{F}(T-1)) = \int_{\mathbb{R}^d} ((m^2+\sigma^2)H(A,y,1) - m^2(H(A,y,1))^2)\psi(dy,T-1).$$

Proof. By the definition of $\psi(A,T)$ we have

$$\mathrm{Var}(\psi(A,T) \mid \mathcal{F}(T-1)) = \sum_{y\in\mathbb{R}^d} \sum_{j=1}^{\lambda(y,T-1)} \mathrm{Var}(I_A(N(y,T-1,j)+y)Z(y,T-1,j)).$$

Clearly

$$\begin{aligned}
&\mathrm{Var}(I_A(N(y,T-1,j)+y)Z(y,T-1,j)) = \\
&= \mathbf{E}(I_A(N(y,T-1,j)+y)Z(y,T-1,j))^2 - \\
&\quad - (\mathbf{E}I_A(N(y,T-1,j)+y)Z(y,T-1,j))^2 = \\
&= (m^2+\sigma^2)H(A,y,1) - m^2(H(A,y,1))^2
\end{aligned}$$

which, in turn, implies the second statement of Lemma 6.5. Observe that

$$\sum_{k=1}^{\infty} \mathrm{Var}(I_{A_k}(N(y,T-1,j)+y)Z(y,T-1,j)) = m^2+\sigma^2 - m^2\sum_{k=1}^{\infty}(H(A_k,y,1))^2.$$

Hence

$$\begin{aligned}
&\sum_{k=1}^{\infty} \mathrm{Var}(\psi(A_k,T) \mid \mathcal{F}(T-1)) = \\
&= \sum_{y\in\mathbb{R}^d} \lambda(y,T-1)(m^2+\sigma^2 - m^2\sum_{k=1}^{\infty}(H(A_k,y,1))^2) \le B(T-1)(m^2+\sigma^2)
\end{aligned}$$

and we have Lemma 6.5.

For any $0 < t < T$ let

$$h(x,y,t) = (2\pi t)^{-d/2} \exp\left(-\frac{\|x-y\|^2}{2t}\right)$$

and

$$f(x,T,t) = m^{T-t}\int_{\mathbb{R}^d} h(x,y,T-t)\psi(dy,t) = m^{T-t}\sum_{y\in\mathbb{R}^d} h(x,y,T-t)\lambda(y,t).$$

Observe that

$$\int_{\mathbb{R}^d} h^2(x,y,t)dx = \int_{\mathbb{R}^d} h(x,y,t)h(y,x,t)dt = h(0,0,2t),$$

$$F(A,T,t) = \int_A f(x,T,t)dx \qquad (0<t<T) \tag{6.6}$$

and by Lemma 6.4 for any $0 < t < T$

$$\mathbf{E}(f(x,T,t) \mid \mathcal{F}(t-1)) = f(x,T,t-1). \tag{6.7}$$

LEMMA 6.6 *Let $\varepsilon > 0$, $\xi \in \mathbb{R}^d$ with $\|\xi\| \geq 3\varepsilon$. Then*

$$\mathbf{E}\int_{\mathbb{R}^d} H(C(\xi T, \varepsilon T), y, 1)\psi(dy, T-1) \leq m^T \exp\left(-\frac{(\|\xi\| - 3\varepsilon)^2}{2}T\right)$$

if T is big enough.

Proof. By Lemma 6.1

$$\begin{aligned}
&\mathbf{E}\int_{\mathbb{R}^d} H(C(\xi T, \varepsilon T), y, 1)\psi(dy, T-1) = \\
&= m^{T-1}\int_{\mathbb{R}^d} H(C(\xi T, \varepsilon T), y, 1)H(dy, 0, T-1) = \\
&= m^{T-1}\int_{\mathbb{R}^d} H(C(\xi T, \varepsilon T), y, 1)h(y, 0, T-1)dy = \\
&= m^{T-1}\int_{C(\xi T, 2\varepsilon T)} H(C(\xi T, \varepsilon T), y, 1)h(y, 0, T-1)dy + \\
&\quad + m^{T-1}\int_{\mathbb{R}^d - C(\xi T, 2\varepsilon T)} H(C(\xi T, \varepsilon T), y, 1)h(y, 0, T-1)dy.
\end{aligned}$$

Observe that

$$h(y, 0, T-1) \leq \exp\left(-\frac{(\|\xi\| - 3\varepsilon)^2 T}{2}\right)$$

if $y \in C(\xi T, 2\varepsilon T)$ and

$$H(C(\xi T, \varepsilon T), y, 1) \leq \exp\left(-\frac{\varepsilon^2 T^2}{3}\right)$$

if $y \notin C(\xi T, 2\varepsilon T)$.

Hence we have Lemma 6.6.

LEMMA 6.7 (cf. Lemma 4.6) *For any $0 < t < T$ we have*

$$\begin{aligned}
&\int_{\mathbb{R}^d} \mathbf{E}((f(x, T, t) - f(x, T, t-1))^2 \mid \mathcal{F}(t-1))dx = \\
&= m^{2T-2t}B(t-1)\left(\frac{m^2 + \sigma^2}{(4\pi(T-t))^{d/2}} - \frac{m^2}{(4\pi(T-t+1))^{d/2}}\right) \leq \\
&\leq m^{2T-2t}B(t-1)\frac{m^2 + \sigma^2}{(4\pi(T-t))^{d/2}}
\end{aligned}$$

and

$$\begin{aligned}
&\mathbf{E}((f(x, T, t) - f(x, T, t-1))^2 \mid \mathcal{F}(t-1)) \leq \\
&\leq m^{2T-2t}\int_{\mathbb{R}^d}\int_{\mathbb{R}^d} h^2(x, y, T-t)h(u, y, 1)dy\psi(du, t-1).
\end{aligned}$$

Proof. By the definition of f and ψ we have

$$f(x,T,t) = m^{T-t}\int_{\mathbb{R}^d} h(x,y,T-t)\sum_{u\in\mathbb{R}^d}\sum_{j=1}^{\lambda(u,t-1)} I_{dy}(N(u,t-1,j)+u)Z(u,t-1,j).$$

Hence

$$\begin{aligned}
&f(x,T,t) - f(x,T,t-1) = \\
&= m^{T-t}\int_{\mathbb{R}^d} h(x,y,T-t)\sum_{u\in\mathbb{R}^d}\sum_{j=1}^{\lambda(u,t-1)} I_{dy}(N(u,t-1,j)+u)Z(u,t-1,j) - \\
&\quad - m^{T-t+1}\sum_{u\in\mathbb{R}^d}\sum_{j=1}^{\lambda(u,t-1)} h(x,u,T-t+1) = m^{T-t}\sum_{u\in\mathbb{R}^d}\sum_{j=1}^{\lambda(u,t-1)} Q(u,x,j,t,T)
\end{aligned}$$

where

$$\begin{aligned}
&Q(u,x,j,t,T) = Q = \\
&= \int_{\mathbb{R}^d} h(x,y,T-t)I_{dy}(N(u,t-1,j)+u)Z(u,t-1,j) - mh(x,u,T-t+1).
\end{aligned}$$

Observe that

$$\begin{aligned}
&\mathbf{E}\int_{\mathbb{R}^d} h(x,y,T-t)I_{dy}(N(u,t-1,j)+u)Z(u,t-1,j) = \\
&= m\int_{\mathbb{R}^d} h(x,y,T-t)h(u,y,1)dy = mh(x,u,T-t+1).
\end{aligned}$$

Hence

$$\begin{aligned}
&\mathbf{E}(Q^2 \mid \mathcal{F}(t-1)) = \mathbf{E}Q^2 = \\
&= (m^2+\sigma^2)\int_{\mathbb{R}^d} h^2(x,y,T-t)h(u,y,1)dy - m^2h^2(x,u,T-t+1)
\end{aligned}$$

which, in turn, implies the second statement of Lemma 6.7. Observe that

$$\begin{aligned}
&\int_{\mathbb{R}^d} \mathbf{E}Q^2dx = \\
&= (m^2+\sigma^2)\int_{\mathbb{R}^d} h(u,y,1)\int_{\mathbb{R}^d} h^2(x,y,T-t)dxdy - m^2\int_{\mathbb{R}^d} h^2(x,u,T-t+1)dx = \\
&= h(0,0,2T-2t)(m^2+\sigma^2)\int_{\mathbb{R}^d} h(u,y,1)dy - m^2h(0,0,2T-2t+2) = \\
&= (m^2+\sigma^2)h(0,0,2T-2t) - m^2h(0,0,2T-2t+2) = \\
&= \frac{m^2+\sigma^2}{(4\pi(T-t))^{d/2}} - \frac{m^2}{(4\pi(T-t+1))^{d/2}}.
\end{aligned}$$

Consequently

$$\mathbf{E}((f(x,T,t) - f(x,T,t-1))^2 \mid \mathcal{F}(t-1)) = m^{2T-2t}\sum_{u\in\mathbb{R}^d}\lambda(u,t-1)\mathbf{E}(Q^2 \mid \mathcal{F}(t-1))$$

and

$$\int_{\mathbb{R}^d} \mathbf{E}((f(x,T,t)-f(x,T,t-1))^2 \mid \mathcal{F}(t-1))dx =$$
$$= m^{2T-2t}B(t-1)\left(\frac{m^2+\sigma^2}{(4\pi(T-t))^{d/2}} - \frac{m^2}{(4\pi(T-t+1))^{d/2}}\right).$$

This proves Lemma 6.7.

LEMMA 6.8 *Let* $0 < t = \mu T < T$ *and* $x \in C(\xi T, \varepsilon T) \subset \mathbb{R}^d$ *where* $\xi \in \mathbb{R}^d$ *and* $0 < 2\varepsilon < \|\xi\|$. *Then*

$$\int_{\mathbb{R}^d}\int_{\mathbb{R}^d} h^2(x,y,T-t)h(u,y,1)dy\psi(du,t-1) \le$$
$$\le B(t-1)\exp\left(-\frac{(\|\xi\|-2\varepsilon)^2T^2}{T+t}\right) \qquad a.s.$$

for all but finitely many t.

Proof. Observe that

$$\psi(C(\xi T,\varepsilon T),t-1) \le \psi(\mathbb{R}^d - C(0,(\|\xi\|-\varepsilon)T),t-1) \le$$
$$\le B(t-1)\exp\left(-\frac{(\|\xi\|-2\varepsilon)^2}{2(t-1)}T^2\right) \qquad \text{a.s.}$$

for all but finitely many t and

$$h(u,y,1) \le e^{-T} \quad \text{if} \quad \|u-y\| > 2T^{1/2}.$$

Hence we consider only the integral

$$\int_{C(0,(\|\xi\|-\varepsilon)T)}\int_{C(u,2T^{1/2})} h^2(x,y,T-t)h(u,y,1)dy\psi(du,t-1).$$

Let $\|u\| = \lambda\|\xi\|T$ $(0 \le \lambda < 1)$. Then

$$\psi(C(u,du),t-1) \le B(t-1)\exp\left(-\frac{\lambda^2(\|\xi\|-\varepsilon)^2}{2(t-1)}T^2\right)$$

and for $x \in C(\xi T,\varepsilon T)$, $y \in C(u,2T^{1/2})$ we have

$$h^2(x,y,T-t) \le \exp\left(-\frac{\|\xi\|^2(1-\lambda)^2}{T-t}T^2\right).$$

Since for any $0 \le \lambda < 1$ and $0 < t = \mu T < T$

$$\frac{\lambda^2(\|\xi\|-\varepsilon)^2}{2(t-1)}T^2 + \frac{\|\xi\|^2(1-\lambda)^2}{T-t}T^2 \ge \frac{(\|\xi\|-2\varepsilon)^2}{T+t}T^2$$

Lemma 6.8 easily follows.

LEMMA 6.9 (cf. Lemmas 4.7 and 4.8). *Let* $0 < t < T - 1$. *Then*

$$\int_{\mathbb{R}^d} \mathbf{E}((f(x,T,T-1) - f(x,T,t))^2 \mid \mathcal{F}(t))dx \le \frac{CB(t)}{(T-t)^{d/2}} m^{2T-2t}$$

with an absolute constant $C > 0$.

Proof. We have

$$\begin{aligned}
&\mathbf{E}((f(x,T,T-1) - f(x,T,t))^2 \mid \mathcal{F}(t)) = \\
&= \mathbf{E}\left(\left(\sum_{j=t+1}^{T-1} f(x,T,j) - f(x,T,j-1) \right)^2 \middle| \mathcal{F}(t) \right) = \\
&= \sum_{j=t+1}^{T-1} \mathbf{E}((f(x,T,j) - f(x,T,j-1))^2 \mid \mathcal{F}(t)) + \\
&\quad + 2 \sum_{t<j<\ell<T} \mathbf{E}((f(x,T,\ell) - f(x,T,\ell-1))(f(x,T,j) - f(x,T,j-1)) \mid \mathcal{F}(t)).
\end{aligned} \tag{6.8}$$

Clearly by (6.7)

$$\begin{aligned}
&\mathbf{E}((f(x,T,\ell) - f(x,T,\ell-1))(f(x,T,j) - f(x,T,j-1)) \mid \mathcal{F}(t)) = \\
&= \mathbf{E}(\mathbf{E}((f(x,T,\ell) - f(x,T,\ell-1))(f(x,T,j) - f(x,T,j-1)) \mid \mathcal{F}(\ell-1)) \mid \mathcal{F}(t)) = 0
\end{aligned} \tag{6.9}$$

and by Lemma 6.7

$$\begin{aligned}
&\int_{\mathbb{R}^d} \mathbf{E}(\mathbf{E}((f(x,T,j) - f(x,T,j-1))^2 \mid \mathcal{F}(j-1)) \mid \mathcal{F}(t))dx \le \\
&\le m^{2T-2j} \mathbf{E}(B(j-1) \mid \mathcal{F}(t)) \frac{m^2+\sigma^2}{(4\pi(T-j))^{d/2}} = m^{2T-j-t-1} B(t) \frac{m^2+\sigma^2}{(4\pi(T-j))^{d/2}}.
\end{aligned}$$

Hence

$$\begin{aligned}
&\int_{\mathbb{R}^d} \mathbf{E}((f(x,T,T-1) - f(x,T,t))^2 \mid \mathcal{F}(t))dx \le \\
&\le \frac{B(t)(m^2+\sigma^2)}{(4\pi)^{d/2}} m^{2T-t-1} \sum_{j=t+1}^{T-1} \frac{m^{-j}}{(T-j)^{d/2}} \le C \frac{B(t)}{(T-t)^{d/2}} m^{2T-2t}
\end{aligned}$$

and we have Lemma 6.9.

LEMMA 6.10 *Let* $0 < t < T$ *and* $x \in C(\xi T, \varepsilon T) \subset \mathbb{R}^d$ *where* $\xi \in \mathbb{R}^d$ *and* $0 < 3\varepsilon < \|\xi\| < (2\log m)^{1/2}$. *Then*

$$\mathbf{E}((f(x,T,T-1) - f(x,T,t))^2 \mid \mathcal{F}(t)) \le B(t) m^{2T-2t} \exp\left(-\frac{(\|\xi\| - 3\varepsilon)^2}{T+t} T^2 \right) \quad a.s.$$

for all but finitely many t.

Proof. By Lemmas 6.7 and 6.8 we have

$$\begin{aligned}
&\mathbf{E}((f(x,T,j)-f(x,T,j-1))^2 \mid \mathcal{F}(t)) = \\
&= \mathbf{E}(\mathbf{E}((f(x,T,j)-f(x,T,j-1))^2 \mid \mathcal{F}(j-1)) \mid \mathcal{F}(t)) \leq \\
&\leq m^{2T-j-t-1}B(t)\exp\left(-\frac{(\|\xi\|-2\varepsilon)^2}{T+j}T^2\right).
\end{aligned} \tag{6.10}$$

By (6.8), (6.9) and (6.10)

$$\begin{aligned}
&\mathbf{E}((f(x,T,T-1)-f(x,T,t))^2 \mid \mathcal{F}(t)) = \\
&= \sum_{j=t+1}^{T-1} \mathbf{E}((f(x,T,j)-f(x,T,j-1))^2 \mid \mathcal{F}(t)) \leq \\
&\leq m^{2T-t-1}B(t)\sum_{j=t+1}^{T-1} m^{-j}\exp\left(-\frac{(\|\xi\|-2\varepsilon)^2}{T+j}T^2\right) \leq \\
&\leq m^{2T-2t}B(t)\exp\left(-\frac{(\|\xi\|-3\varepsilon)^2}{T+t}T^2\right).
\end{aligned}$$

Hence we have Lemma 6.10.

LEMMA 6.11 (cf. Lemma 4.8) *Let $A_1, A_2, \ldots$ be a partition of $\mathbb{R}^d$ with $|A_k| \leq K$ $(k = 1, 2, \ldots,\ K > 0)$. Then for any $0 < t < T$ we have*

$$\mathbf{E}\left(\sum_{k=1}^{\infty}\left(\frac{\psi(A_k,T)}{m^T} - \frac{F(A_k,T,t)}{m^T}\right)^2 \,\middle|\, \mathcal{F}(t)\right) \leq C\frac{B(t)}{m^{2t}(T-t)^{d/2}} \tag{6.11}$$

where $C = C(m,d,K) > 0$ is a constant.

Proof. Observe that by Lemma 6.1

$$\begin{aligned}
&\mathbf{E}((\psi(A,T)-F(A,T,t))^2 \mid \mathcal{F}(t)) = \\
&= \mathbf{E}((\psi(A,T)-F(A,T,T-1)+F(A,T,T-1)-F(A,T,t))^2 \mid \mathcal{F}(t)) = \\
&= \mathbf{E}(\mathbf{E}((\psi(A,T)-F(A,T,T-1))^2 \mid \mathcal{F}(T-1)) \mid \mathcal{F}(t)) + \\
&\quad + \mathbf{E}((F(A,T,T-1)-F(A,T,t))^2 \mid \mathcal{F}(t)).
\end{aligned} \tag{6.12}$$

By Lemmas 6.1 and 6.5 we have

$$\sum_{k=1}^{\infty}\mathbf{E}(\mathbf{E}((\psi(A_k,T)-F(A_k,T,T-1))^2 \mid \mathcal{F}(T-1)) \mid \mathcal{F}(t)) \leq (m^2+\sigma^2)m^{T-t-1}B(t). \tag{6.13}$$

The Cauchy inequality, (6.6) and Lemma 6.9 imply

$$\sum_{k=1}^{\infty}\mathbf{E}((F(A_k,T,T-1)-F(A_k,T,t))^2 \mid \mathcal{F}(t)) =$$
$$= \sum_{k=1}^{\infty}\mathbf{E}\left(\left(\int_{A_k}(f(x,T,T-1)-f(x,T,t))dx\right)^2 \,\middle|\, \mathcal{F}(t)\right) \le$$
$$\le \sum_{k=1}^{\infty}\mathbf{E}\left(|A_k|\int_{A_k}(f(x,T,T-1)-f(x,T,t))^2dx \,\middle|\, \mathcal{F}(t)\right) \le$$
$$\le K\int_{\mathbb{R}^d}\mathbf{E}((f(x,T,T-1)-f(x,T,t))^2 \mid \mathcal{F}(t))dx \le \frac{CB(t)}{(T-t)^{d/2}}m^{2T-2t}. \quad (6.14)$$

Now, we have (6.11) by (6.12), (6.13) and (6.14).

LEMMA 6.12 *Let $0 \le t < T$, $\xi \in \mathbb{R}^d$ with $0 < 3\varepsilon < \|\xi\| < (2\log m)^{1/2}$ and L is a big enough positive constant depending only on d. Then*

$$\mathbf{E}(\psi(C(\xi T,\varepsilon T),T)-F(C(\xi T,\varepsilon T),T,t))^2 \le$$
$$\le L\left(m^{2T-t}\exp\left(-\frac{(\|\xi\|-3\varepsilon)^2}{T+t}T^2\right)+m^T\exp\left(-\frac{(\|\xi\|-3\varepsilon)^2}{2}T\right)\right)$$

if T is big enough.
Consequently

$$\mathbf{E}|\psi(C(\xi T,\varepsilon T),T)-F(C(\xi T,\varepsilon T),T,t)| \le Lm^{T-t/2}\exp\left(-\frac{(\|\xi\|-3\varepsilon)^2}{2(T+t)}T^2\right).$$

Proof. By (6.12) we have

$$\mathbf{E}(\psi(A,T)-F(A,T,t))^2 = \mathbf{E}(\mathbf{E}((\psi(A,T)-F(A,T,t))^2 \mid \mathcal{F}(t))) =$$
$$= \mathbf{E}(\psi(A,T)-F(A,T,T-1))^2 + \mathbf{E}(\mathbf{E}((F(A,T,T-1)-F(A,T,t))^2 \mid \mathcal{F}(t))).$$

By Lemmas 6.1, 6.5 and 6.6

$$\mathbf{E}(\psi(C(\xi T,\varepsilon T),T)-F(C(\xi T,\varepsilon T),T,T-1))^2 =$$
$$= \mathbf{E}(\mathbf{E}((\psi(C(\xi T,\varepsilon T),T)-\mathbf{E}(\psi(C(\xi T,\varepsilon T),T) \mid \mathcal{F}(T-1)))^2 \mid \mathcal{F}(T-1))) \le$$
$$\le (m^2+\sigma^2)\int_{\mathbb{R}^d}H(C(\xi T,\varepsilon T),y,1)\mathbf{E}\psi(dy,T-1) \le$$
$$\le (m^2+\sigma^2)m^T\exp\left(-\frac{(\|\xi\|-3\varepsilon)^2}{2}T\right).$$

By (6.6), the Cauchy inequality and Lemma 6.10

$$\mathbf{E}((F(C(\xi T,\varepsilon T),T,T-1)-F(C(\xi T,\varepsilon T),T,t))^2 \mid \mathcal{F}(t)) =$$

$$= \mathbf{E}\left(\left(\int_{C(\xi T,\varepsilon T)}(f(x,T,T-1)-f(x,T,t))dx\right)^2 \,\middle|\, \mathcal{F}(t)\right) \le$$
$$\le K(\varepsilon T)^{d/2}\int_{C(\xi T,\varepsilon T)}\mathbf{E}((f(x,T,T-1)-f(x,T,t))^2 \mid \mathcal{F}(t))dx \le$$
$$\le K(\varepsilon T)^{d/2}\sup_{x\in C(\xi T,\varepsilon T)}\mathbf{E}((f(x,T,T-1)-f(x,T,t))^2 \mid \mathcal{F}(t)) \le$$
$$\le K(\varepsilon T)^{d/2}B(t)m^{2T-2t}\exp\left(-\frac{(\|\xi\|-2\varepsilon)^2}{T+t}T^2\right).$$

Hence we have Lemma 6.12.

THEOREM 6.1 (cf. Theorem 4.3) *Let $A_1, A_2, \ldots$ be a partition of $C(0,T^\alpha)$ $(0 \le \alpha \le 1)$ with $0 < K_1 \le |A_k| \le K_2 < \infty$ $(k = 1,2,\ldots)$. Then for any $0 < t < T$ we have*

$$\mathbf{E}\left(\sum_{k=1}^{\infty}\left|\frac{\psi(A_k,T)}{m^T}-\frac{F(A_k,T,t)}{m^T}\right| \,\middle|\, \mathcal{F}(t)\right) \le CT^{\alpha d/2}\left(\frac{1}{m^t(T-t)^{d/2}}\frac{B(t)}{m^t}\right)^{1/2} \tag{6.15}$$

and

$$\mathbf{E}\sum_{k=1}^{\infty}\left|\frac{\psi(A_k,T)}{m^T}-\frac{F(A_k,T,t)}{m^T}\right| \le CT^{\alpha d/2}(m^t(T-t)^{d/2})^{-1/2} \tag{6.16}$$

with a constant $C = C(m,d,K_1,K_2) > 0$.

Proof. By Lemma 6.11 and the Cauchy inequality we have

$$\mathbf{E}\left(\sum_{k=1}^{\infty}\left|\frac{\psi(A_k,T)}{m^T}-\frac{F(A_k,T,t)}{m^T}\right| \,\middle|\, \mathcal{F}(t)\right) \le$$
$$\le \left|\frac{C(0,T^\alpha)}{K_1}\right|^{1/2}\left(\mathbf{E}\sum_{k=1}^{\infty}\left(\frac{\psi(A_k,T)}{m^T}-\frac{F(A_k,T,t)}{m^T}\right)^2 \,\middle|\, \mathcal{F}(t)\right)^{1/2} \le$$
$$\le CT^{\alpha d/2}\left(\frac{1}{m^t(T-t)^{d/2}}\frac{B(t)}{m^t}\right)^{1/2}.$$

Hence we have (6.15). Since $\mathbf{E}(m^{-t}B(t))^{1/2} \le 1$, (6.15) implies (6.16).

6.3 Global limit theorems

Our main result runs as follows:

THEOREM 6.2 (cf. Theorem 4.4) *Let $A_1, A_2, \ldots$ be a partition of $C(0,T)$ with*

$$0 < K_1 \le |A_k| \le K_2 < \infty$$

and

$$0 < \operatorname{diam} A_k = \sup_{x,y \in A_k} \|x - y\| \le K_3 < \infty \qquad (k = 1, 2, \ldots).$$

Then for any $0 < \varepsilon < 1/2$ there exists a $C = C(\varepsilon) > 0$ such that for any $T = 1, 2, \ldots$ we have

$$\mathbf{E} \sum_{k=1}^{\infty} \left| \frac{\psi(A_k, T)}{m^T} - H(A_k, 0, T) B \right| \le C T^{-(1/2-\varepsilon)}.$$

The proof of Lemma 4.9 was based on the trivial fact that

$$\lambda(x, t) = 0 \qquad \text{if} \qquad |x| > t$$

in case of the branching random walk. Since it is not true for branching Wiener process, at first we prove our

LEMMA 6.13 *Let*

$$K > (2 \log m)^{1/2}.$$

Then

$$\mathbf{P}\{\psi(\mathbb{R}^d - \mathcal{C}(0, Kt), t) \ne 0\} \le e^{-Ct}$$

with some $C > 0$ and

$$\psi(\mathbb{R}^d - \mathcal{C}(0, Kt), t) = 0 \qquad a.s.$$

for all but finitely many t.

Proof. By Theorem 4.2 for any $\varepsilon > 0$

$$B(t) \le m^{(1+\varepsilon)t} \qquad \text{a.s.}$$

for all but finitely many t.

Let $W(t) \in \mathbb{R}^d$ $(t \ge 0)$ be a Wiener process. Then

$$\mathbf{P}\{W(t) \notin \mathcal{C}(0, Kt)\} \le C \exp\left(-\frac{K^2}{2} t\right)$$

with some $C > 0$.

Hence

$$\mathbf{P}\{\psi(\mathbb{R}^d - \mathcal{C}(0, Kt), t) \ne 0\} \le C m^{(1+\varepsilon)t} \exp\left(-\frac{K^2}{2} t\right)$$

with some $C > 0$ which, in turn, proves Lemma 6.13.

LEMMA 6.14 (cf. Lemma 4.9) *Let $0 < t < T$ and A be a Borel set of $\mathbb{R}^d$. Then*

$$\mathbf{E}(\psi(A,T) \mid \mathcal{F}(t)) \geq m^T \frac{B(t)}{m^t} \inf_{y \in C(0,Kt)} H(A,y,T-t)$$

and

$$\mathbf{E}(\psi(A,T) \mid \mathcal{F}(t)) \leq m^T \frac{B(t)}{m^t} \sup_{y \in C(0,Kt)} H(A,y,T-t)$$

almost surely for all but finitely many t provided that

$$K > (2 \log m)^{1/2}.$$

Proof. Let

$$\Omega_1 = \Omega_1(t,K) = \{\omega : \psi(\mathbb{R}^d - C(0,Kt), t) = 0\}.$$

Then on Ω_1

$$\mathbf{E}(\psi(A,T) \mid \mathcal{F}(t)) \geq \inf_{y \in C(0,Kt)} \mathbf{E}(\psi(A,T) \mid \lambda(y,t) = B(t)) =$$
$$= m^T \frac{B(t)}{m^t} \inf_{y \in C(0,Kt)} H(A,y,T-t)$$

and

$$\mathbf{E}(\psi(A,T) \mid \mathcal{F}(t)) \leq \sup_{y \in C(0,Kt)} \mathbf{E}(\psi(A,T) \mid \lambda(y,t) = B(t)) =$$
$$= m^T \frac{B(t)}{m^t} \sup_{y \in C(0,Kt)} H(A,y,T-t).$$

Hence we have Lemma 6.14 by Lemma 6.13.

LEMMA 6.15 (cf. Lemma 4.10) *Let $K > 0$ and assume that*

$$\|y\| \leq Kt, \qquad \|x\| \leq T, \qquad 0 < t \leq T/2.$$

Then we have

$$\left| \frac{h(x,0,T)}{h(x,y,T-t)} - 1 \right| \leq C \frac{\|x\|t + t^2}{T}$$

with some constant $C = C(K) > 0$.

Let $A \subset \mathbb{R}^d$ be a Borel set with $0 < K_1 \leq |A| \leq K_2 < \infty$ and diam $A \leq K_3 < \infty$. *Then*

$$1 - C \int_A \frac{\|x\|t + t^2}{T} dx \leq \frac{H(A,0,T)}{\sup_{y \in C(0,Kt)} H(A,y,T-t)} \leq$$
$$\leq \frac{H(A,0,T)}{\inf_{y \in C(0,Kt)} H(A,y,T-t)} \leq 1 + C \int_A \frac{\|x\|t + t^2}{T} dx$$

where $C = C(K_1, K_2, K_3) > 0$. Further for any $\alpha > 1/2$

$$H(\mathbb{R}^d - C(0,T^\alpha), 0, T) \leq \exp(-O(T^{2\alpha-1})). \tag{6.17}$$

Proof Observe that

$$\frac{h(x,0,T)}{h(x,y,T-t)} = \left(1-\frac{t}{T}\right)^{d/2} \exp\left(-\frac{1}{2}\left(\frac{\|x\|^2}{T} - \frac{\|y-x\|^2}{T-t}\right)\right)$$

and

$$\begin{aligned}\left|\frac{\|x\|^2}{T} - \frac{\|y-x\|^2}{T-t}\right| &= \left|\frac{\|x\|^2(T-t) - (\|y\|^2 + \|x\|^2 - 2(x,y))T}{T(T-t)}\right| \le \\ &\le \frac{\|x\|^2 t + \|y\|^2 T + 2T\|x\|\|y\|}{T(T-t)} \le \\ &\le \frac{2(\|x\|tT + K^2t^2T + 2KTt\|x\|)}{T^2} \le \\ &\le \frac{2(\|x\|t + K^2t^2 + 2Kt\|x\|)}{T}.\end{aligned}$$

Hence we have the first statement of Lemma 6.15.

In order to prove its second statement observe that

$$\begin{aligned}&|H(A,0,T) - H(A,y,T-t)| \le \\ &\le C\int_A h(x,y,T-t)\frac{\|x\|t+t^2}{T}dx \le C\sup_{x\in A} h(x,y,T-t)\int_A \frac{\|x\|t+t^2}{T}dx \le \\ &\le \frac{C}{|A|}H(A,y,T-t)\frac{|A|\sup_{x\in A}h(x,y,T-t)}{H(A,y,T-t)}\int_A \frac{\|x\|t+t^2}{T}dx.\end{aligned}$$

Since

$$\frac{|A|\sup_{x\in A}h(x,y,T-t)}{H(A,y,T-t)} \le \exp\left(-C\frac{K_3\|x\|}{T-t}\right),$$

we have our second statement.

(6.17) is trivial. Hence Lemma 6.15 is proved.

Proof of Theorem 6.2. For any $0 < t < T$ and $0 < \alpha < 1$ we have

$$\begin{aligned}&\sum_{k=1}^{\infty}\left|\frac{\psi(A_k,T)}{m^T} - H(A_k,0,T)B\right| \le \\ &\le \frac{\psi(\mathbb{R}^d - C(0,T^\alpha),T)}{m^T} + H(\mathbb{R}^d - C(0,T^\alpha),0,T)B + \\ &\quad + \sum_{k=1}^{\infty}\left|\frac{\psi(A_kC(0,T^\alpha),T)}{m^T} - \frac{F(A_kC(0,T^\alpha),T,t)}{m^T}\right| + \\ &\quad + \sum_{k=1}^{\infty}\left|\frac{F(A_kC(0,T^\alpha),T,t)}{m^T} - H(A_kC(0,T^\alpha),0,T)\frac{B(t)}{m^t}\right| + \\ &\quad + \sum_{k=1}^{\infty}\left|H(A_kC(0,T^\alpha),0,T)\frac{B(t)}{m^t} - H(A_kC(0,T^\alpha),0,T)B\right|. \qquad (6.18)\end{aligned}$$

Choosing $\alpha > 1/2$ by (6.17) and (6.2) we have

$$H(\mathbb{R}^d - C(0,T^\alpha),0,T) \le \exp(-O(T^{2\alpha-1})) \tag{6.19}$$

$$\mathbf{E}\frac{\psi(\mathbb{R}^d - C(0,T^\alpha),T)}{m^T} \le \exp(-O(T^{2\alpha-1})). \tag{6.20}$$

Let $T = [t^\beta]$ with $\beta > 1$ by Theorem 6.1

$$m^{-T}\sum_{k=1}^{\infty}\mathbf{E}|\psi(A_k C(0,T^\alpha),T) - F(A_k, C(0,t^\alpha),T,t)| \le$$

$$\le CT^{\alpha d/2}(m^t(T-t)^{d/2})^{-1/2} \le \exp(-O(T^{1/\beta})). \tag{6.21}$$

Choosing $\alpha > 1/2$ close enough to $1/2$ and β big enough by Lemmas 6.14 and 6.15 on Ω_1 (cf. Lemma 6.3) we have

$$\sum_{k=1}^{\infty}\left|\frac{F(A_k C(0,T^\alpha),T,t)}{m^T} - H(A_k C(0,T^\alpha),0,T)\frac{B(t)}{m^t}\right| \le$$

$$\le \frac{B(t)}{m^t}O(1)\sum_{k=1}^{\infty}H(A_k C(0,T^\alpha),0,T)\int_{A_k C(0,T^\alpha)}\frac{\|x\|t+t^2}{T}dx \le$$

$$\le \frac{B(t)}{m^t}O(1)\sum_{k=1}^{\infty}T^{-d/2}\int_{A_k C(0,T^\alpha)}\frac{\|x\|t+t^2}{T}dx =$$

$$= \frac{B(t)}{m^t}O(1)T^{-(d/2+1)}\int_{C(0,T^\alpha)}(\|x\|t+t^2)dx = O(T^{-(1/2-\varepsilon)}) \quad \text{a.s.} \tag{6.22}$$

for any $\varepsilon > 0$ and for all but finitely many t.

By Lemma 6.13

$$\int_{\Omega-\Omega_1}\frac{B(t)}{m^t}d\mathbf{P} \le \left(\int_\Omega \frac{B^2(t)}{m^{2t}}d\mathbf{P}\,\mathbf{P}(\Omega-\Omega_1)\right)^{1/2} \le e^{-Ct}, \tag{6.23}$$

$$m^{-T}\int_{\Omega-\Omega_1}F(A_k C(0,T^\alpha),T,t)d\mathbf{P} \le m^{-T}\int_{\Omega-\Omega_1}B(t)m^{T-t}d\mathbf{P} \le e^{-Ct}. \tag{6.24}$$

(6.20), (6.21) and (6.22) imply

$$\mathbf{E}\sum_{k=1}^{\infty}\left|\frac{F(A_k C(0,T^\alpha),T,t)}{m^T} - H(A_k C(0,T^\alpha),0,T)\frac{B(t)}{m^t}\right| \le O(T^{1/2-\varepsilon}). \tag{6.25}$$

By (4.3)

$$\mathbf{E}\sum_{k=1}^{\infty}\left|H(A_k C(0,T^\alpha),0,T)\frac{B(t)}{m^t} - H(A_k C(0,T^\alpha),0,T)B\right| \le \exp(-O(T^{1/\beta})). \tag{6.26}$$

(6.18), (6.19) (6.20), (6.21), (6.25) and (6.26) imply Theorem 6.2.

THEOREM 6.3 (cf. Theorem 4.5) *Let $A_1, A_2, \ldots$ be a partition of $C(0,T)$ with $0 < K_1 \le |A_k| \le K_2 < \infty$, $\operatorname{diam} A_k \le K_3 < \infty$ $(k = 1,2,\ldots)$. Then for any $0 < \varepsilon < 1/2$*

$$\lim_{T\to\infty} T^{1/2-\varepsilon} \sum_{k=1}^{\infty} \left| \frac{\psi(A_k,T)}{m^T} - H(A_k,0,T)B \right| = 0 \qquad a.s.$$

where the partition $\{A_k\}$ might depend on T.

Proof is the same as that of Theorem 4.5.

Theorem 6.3, Lemma 6.13 and (6.17) easily imply

THEOREM 6.4 (cf. Theorem 4.6) *For any $x \in I\!R^d$ and $\varepsilon > 0$ we have*

$$\lim_{T\to\infty} T^{1/2-\varepsilon} \left| \frac{\psi(\{y:\ y \le xT^{1/2}\},T)}{m^T} - B\Phi(x) \right| = 0 \qquad a.s.$$

6.4 Local limit theorems

The two theorems of this Section can be proved simply copying the proofs of Theorems 4.7 and 4.8

THEOREM 6.5 (cf. Theorem 4.7) *Let $A_1, A_2, \ldots$ be a partition of $C(0,T^\gamma)$ $(0 \le \gamma \le 1/2)$ with $0 < K_1 \le |A_k| \le K_2 < \infty$, $\operatorname{diam} A_k \le K_3 < \infty$ $(k = 1,2,\ldots)$. Then for any $\varepsilon > 0$ we have*

$$\mathbf{E}\left(\sum_{k=1}^{\infty} \left| \frac{\psi(A_k,T)}{m^T} - H(A_k,0,T)B \right| \right) \le CT^{-(d+2-2\gamma(d+1)-\varepsilon)/2}$$

and

$$\lim_{T\to\infty} T^{(d+2-2\gamma(d+1)-\varepsilon)/2} \sum_{k=1}^{\infty} \left| \frac{\psi(A_k,T)}{m^T} - H(A_k,0,T)B \right| = 0 \qquad a.s.$$

THEOREM 6.6 (cf. Theorem 4.8) *Let $A = A(T) \subset C(0,T^\gamma)$ $(0 \le \gamma \le 1)$ be a sequence of Borel sets with $0 < K_1 \le |A(T)| \le K_2 < \infty$, $\operatorname{diam} A(T) \le K_3 < \infty$ $(T = 1,2,\ldots)$. Then for any $0 < \varepsilon < \gamma$*

$$\mathbf{P}\left\{ T^{(d+2-2\gamma-2\varepsilon)/2} \left| \frac{\psi(A,T)}{m^T} - H(A,0,T)B \right| \ge 1 \right\} \le \exp(-O(T^\varepsilon))$$

and

$$\lim_{T\to\infty} T^{d+2-2\gamma-\varepsilon} \mathbf{E}\left(\frac{\psi(A,T)}{m^T} - H(A,0,T)B \right)^2 = 0.$$

Consequently for any fixed $A \subset I\!R^d$ $(0 < |A| < \infty)$ and $0 < \varepsilon < 1$ we have

$$\lim_{T\to\infty} T^{1-\varepsilon} \left| \frac{\psi(A,T)}{m^T H(A,0,T)} - B \right| = 0 \qquad a.s.$$

6.5 Particles located far away from the origin

Having Lemma 6.2 we can expect that

$$\lim_{T\to\infty} \psi(C(\xi T, \varepsilon T), T) = \infty \qquad \text{a.s.} \tag{6.27}$$

for any $\varepsilon > 0$ and $\xi \in \mathbb{R}^d$ if $\|\xi\| < (2\log m)^{1/2}$ and

$$\lim_{T\to\infty} \psi(C(\xi T, \varepsilon T), T) = 0 \qquad \text{a.s.} \tag{6.28}$$

if $\|\xi\| > (2\log m)^{1/2}$.

It is really so. In fact we have

THEOREM 6.7 *For any $\varepsilon > 0$ we have*

$$\mathbf{P}\{\psi(\mathbb{R}^d - C(0, ((2\log m)^{1/2} + \varepsilon)T), T) > 0\} \leq e^{-\varepsilon T}.$$

Proof is trivial.

Clearly Theorem 6.7 implies (6.28).

A much stronger version of (6.27) will be proved later on. See Theorem 11.5.

6.6 Generalizations

The generalizations given in Chapter 4 for branching random walk can be given in the same way for branching Wiener processes. We omit the details.

However we are interested in the question whether the simple, symmetric random walk in Chapter 4 resp. the Wiener process in the present Chapter can be replaced by some other random walks. Studying the above given proofs it turns out that the only condition used, is the large deviation principle. In fact, let $Y_1, Y_2, \ldots$ be a sequence of i.i.d. $\mathbb{R}^d$ valued random vectors with

$$\mathbf{E}Y_i = 0, \qquad \mathbf{E}\|Y_i\|^2 = 1, \qquad T_0 = 0, \qquad T_n = Y_1 + Y_2 + \cdots + Y_n$$

$$\mathbf{P}\left\{\frac{\|T_n\|}{n} > a\right\} \leq \exp(-Cn)$$

with some $C = C(a) > 0$.

Having this condition for the random walk, the results of Sections 6.2, 6.3 and 6.4 remain true.

Chapter 7

Critical branching random walk starting with one particle

7.1 Critical branching process

We use the same notations and conditions as in Section 4.1. However in the present Chapter we are interested in the critical case i.e. we assume that

$$m = 1.$$

At first we recall a theorem on critical branching processes.

THEOREM 7.1 ([6] Theorem 1 p. 19, (3) p. 20, Theorem 2 p. 20 and Theorem 1 p. 61.) *As $t \to \infty$ we have*

$$\mathbf{E}B(t) = 1, \quad \mathbf{P}\{B(t) > 0\} \sim \frac{2}{t\sigma^2}, \tag{7.1}$$

$$\mathbf{E}(B(t) \mid B(t) > 0) = \frac{1}{\mathbf{P}\{B(t) > 0\}} \sim \frac{t\sigma^2}{2}, \tag{7.2}$$

$$\mathbf{P}\left\{\frac{B(t)}{t} > z \mid B(t) > 0\right\} \to \exp\left(-\frac{2z}{\sigma^2}\right) \qquad (z \geq 0). \tag{7.3}$$

On the set $\{B(T) > 0\}$ for each fixed $0 < t < 1$

$$\frac{B(tT)}{T} \xrightarrow{\mathcal{D}} U + V \quad as \quad T \to \infty \tag{7.4}$$

where U and V are independent r.v.'s having exponential distributions with parameters $2\sigma^{-2}t^{-1}$ and $2(t(1-t)\sigma^2)^{-1}$ respectively.

(7.3) suggests that

$$\mathbf{P}\{B(t) > zt\log t \mid B(t) > 0\} \sim t^{-2z/\sigma^2}.$$

Instead of proving this asymptotic relation we prove the following much weaker

LEMMA 7.1 *For any $\varepsilon > 0$, $\alpha > 2$ and t big enough we have*

$$\mathbf{P}\{B(t) \geq t^{\alpha} \mid B(t) > 0\} \leq \frac{(1+\varepsilon)\sigma^2}{2t^{\alpha-1}}, \tag{7.5}$$

$$\mathbf{P}\{\max_{0\leq t\leq T} B(t) \geq T^{\alpha} \mid B(T) > 0\} \leq \frac{(1+\varepsilon)\sigma^2}{2T^{\alpha-2}}, \tag{7.6}$$

$$\mathbf{P}\{\max_{0\leq t\leq T} B(t) \geq T^{\alpha} \mid B(T) = 0\} \leq \frac{(1+\varepsilon)\sigma^2}{2T^{\alpha-2}}. \tag{7.7}$$

Proof. (7.5) is a trivial consequence of (7.2) and the Markov inequality.

Let $0 \leq s < t$. Then

$$\begin{aligned}
&\mathbf{E}(B(s) \mid B(t) > 0) = \\
&= \frac{1}{\mathbf{P}\{B(t) > 0\}} \int_{\{B(t)>0\}} B(s) d\mathbf{P} \leq \frac{1}{\mathbf{P}\{B(t) > 0\}} \int_{\{B(s)>0\}} B(s) d\mathbf{P} = \\
&= \frac{\mathbf{P}\{B(s) > 0\}}{\mathbf{P}\{B(t) > 0\}} \mathbf{E}\{B(s) \mid B(s) > 0\} = \frac{1}{\mathbf{P}\{B(t) > 0\}} \sim \frac{t\sigma^2}{2}.
\end{aligned}$$

Hence we have (7.6) by Markov inequality.

Since

$$\mathbf{P}\{\max_{0\leq t\leq T} B(t) \geq T^{\alpha} \mid B(T) > 0\} \geq \mathbf{P}\{\max_{0\leq t\leq T} B(t) \geq T^{\alpha} \mid B(T) = 0\},$$

(7.7) follows from (7.6).

Now we present a few further properties of $B(\cdot)$.

Assuming that

$$\mathbf{E}(B(1))^3 = \sum_{i=0}^{\infty} i^3 p_i \tag{7.8}$$

and following the proof of (7.1) given in [6] we get

LEMMA 7.2

$$\mathbf{P}\{B(t) > 0\} = \frac{2}{t\sigma^2}\left(1 + O\left(\frac{\log t}{t}\right)\right) \tag{7.9}$$

and

$$\mathbf{E}(B(t) \mid B(t) > 0) = \frac{t\sigma^2}{2}\left(1 + O\left(\frac{\log t}{t}\right)\right). \tag{7.10}$$

From now on for sake of simplicity we assume that (7.8) holds true.

LEMMA 7.3

$$\mathbf{E}B^2(t) = \sigma^2 t + 1, \tag{7.11}$$

$$\mathbf{E}(B^2(t) \mid B(t) > 0) = \frac{\sigma^2 t}{2}(\sigma^2 t + 1)\left(1 + O\left(\frac{\log t}{t}\right)\right). \tag{7.12}$$

For any $\varepsilon > 0$, $\alpha > 1$ and t big enough we have

$$\mathbf{P}\{B(t) \geq t^\alpha \mid B(t) > 0\} \leq \frac{(1+\varepsilon)\sigma^4}{2t^{2\alpha-2}}. \tag{7.13}$$

Assume that the μ-th ($\mu = 3, 4, \ldots$) moment of $B(1)$ exists i.e.

$$m_\mu = \mathbf{E}(B(1))^\mu = \sum_{k=0}^{\infty} k^\mu p_k < \infty.$$

Then there exists a $C = C(m_\mu) > 0$ such that

$$\mathbf{P}\{B(t) \geq t^\alpha \mid B(t) > 0\} \leq Ct^{-\mu(\alpha-1)}$$

for any $\alpha > 1$.

Proof. Let $T_1, T_2, \ldots$ be a sequence of i.i.d.r.v.'s and let ν be a nonnegative, integer valued r.v. independent from $\{T_i,\ i = 1, 2, \ldots\}$ with

$$\mathbf{E}T_i = a, \qquad \mathbf{E}T_i^2 = \alpha,$$
$$\mathbf{E}\nu = b, \qquad \mathbf{E}\nu^2 = \beta.$$

Then it is easy to see that

$$\mathbf{E}(T_1 + T_2 + \cdots + T_\nu)^2 = b(\alpha - a^2) + a^2\beta. \tag{7.14}$$

Applying (7.14) and the definition of $B(t)$ in Section 4.1 by induction we get (7.11).

Since by (7.11) and (7.9)

$$\mathbf{E}(B^2(t) \mid B(t) > 0) = \frac{\mathbf{E}B^2(t)}{\mathbf{P}\{B(t) > 0\}} = \frac{\sigma^2 t}{2}(\sigma^2 t + 1)\left(1 + O\left(\frac{\log t}{t}\right)\right),$$

we have (7.12). Since

$$\mathrm{Var}\,(B(t) \mid B(t) > 0) = \left(\frac{\sigma^4 t^2}{4} + \frac{\sigma^2 t}{2}\right)\left(1 + O\left(\frac{\log t}{t}\right)\right) \sim \frac{\sigma^4 t^2}{4},$$

(7.13) follows from the Chebyshev inequality.

The proof of the last inequality of Lemma 7.3 is the same as that of (7.13) by the natural modification of (7.14).

Note that (7.13) is a much stronger statement than (7.5).

For any $0 \leq s < t$ let $Q(s,t)$ be the number of those particles which are living at time s and which have at least one offspring living at time t. Clearly

$$B(s) \geq Q(s,t), \qquad B(t) \geq Q(s,t),$$
$$\{Q(s,t) = 0\} = \{B(t) = 0\} \qquad (0 \leq s \leq t).$$

Now we prove our

LEMMA 7.4 *Assume that $t \to \infty$ and $t - s \to \infty$. Then*

$$\mathbf{E}(Q(s,t) \mid B(t) > 0) \sim \frac{t}{t-s}, \tag{7.15}$$

$$\mathbf{E}(Q^2(s,t) \mid B(t) > 0) \sim \frac{t(t+s)}{(t-s)^2}. \tag{7.16}$$

Proof. By (7.2)

$$\mathbf{E}(B(t) \mid Q(s,t)) \sim \frac{\sigma^2}{2}(t-s)Q(s,t).$$

Hence by (7.1)

$$1 = \mathbf{E}B(t) \sim \frac{\sigma^2}{2}(t-s)\mathbf{E}Q(s,t)$$

and

$$\mathbf{E}Q(s,t) \sim \frac{2}{\sigma^2(t-s)}. \tag{7.17}$$

Clearly

$$\begin{aligned}\mathbf{E}(Q(s,t) \mid B(t) > 0) &= \frac{1}{\mathbf{P}\{B(t) > 0\}} \int_{\{B(t)>0\}} Q(s,t)d\mathbf{P} = \\ &= \frac{1}{\mathbf{P}\{B(t) > 0\}} \int_{\{Q(s,t)>0\}} Q(s,t)d\mathbf{P} = \frac{\mathbf{E}Q(s,t)}{\mathbf{P}\{B(t) > 0\}}\end{aligned}$$

and we have (7.15) by (7.1) and (7.17).

Consider a particle which lives at s and which has at least one offspring living at t. Let $T = T(s,t)$ be the number of offsprings of this particle living at time $t > s$. Observe that by (7.2)

$$\mathbf{E}T \sim \frac{\sigma^2(t-s)}{2},$$

by (7.12)

$$\mathbf{E}T^2 \sim \frac{\sigma^2(t-s)}{2}(\sigma^2(t-s)+1) \quad \text{as} \quad t-s \to \infty$$

and

$$B(t) = T_1 + T_2 + \cdots + T_{Q(s,t)}$$

where $T_1, T_2, \ldots$ are i.i.d.r.v.'s with

$$T_i \stackrel{\mathcal{D}}{=} T(s,t).$$

Hence by (7.12) and (7.14)

$$\mathbf{E}(B^2(t) \mid B(t) > 0) = b(\alpha - a^2) + a^2\beta \sim \frac{\sigma^2 t}{2}(\sigma^2 t + 1)$$

where

$$b = \mathbf{E}(Q(s,t) \mid B(t) > 0) \sim \frac{t}{t-s},$$
$$a = \mathbf{E}T \sim \frac{\sigma^2(t-s)}{2},$$
$$\alpha = \mathbf{E}T^2 \sim \frac{\sigma^2(t-s)}{2}(\sigma^2(t-s)+1),$$
$$\beta = \mathbf{E}(Q^2(s,t) \mid B(t) > 0).$$

Hence we have

$$\mathbf{E}(Q^2(s,t) \mid B(t) > 0) = a^{-2}\left[\frac{\sigma^2 t}{2}(\sigma^2 t + 1) - b(\alpha - a^2)\right]$$

and in turn (7.16) is proved.

LEMMA 7.5 *For any fixed $k = 1, 2, \ldots$ as $n \to \infty$ we have*

$$\mathbf{P}\{B(1) = k \mid B(n) > 0\} \sim \frac{\sigma^2 n}{2}\left[1 - \left(1 - \frac{2}{\sigma^2(n-1)}\right)^k\right] p_k \sim k p_k.$$

Consequently

$$\mathbf{E}(B(1) \mid B(n) > 0) \sim \sum_{k=1}^{\infty} k^2 p_k = \sigma^2 + 1.$$

Proof. Clearly

$$\mathbf{P}\{B(1) = k \mid B(n) > 0\} = \frac{\mathbf{P}\{B(n) > 0 \mid B(1) = k\}\mathbf{P}\{B(1) = k\}}{\mathbf{P}\{B(n) > 0\}}.$$

Since

$$\mathbf{P}\{B(1) = k\} = p_k,$$
$$\mathbf{P}\{B(n) > 0\} \sim \frac{2}{\sigma^2 n},$$
$$\mathbf{P}\{B(n) > 0 \mid B(1) = k\} = 1 - \left(1 - \frac{2}{\sigma^2(n-1)}\right)^k,$$

we have Lemma 7.5.

Note that $Q(s,t)$ is clearly a nondecreasing function of s $(0 \le s \le t)$ and $Q(0,t) = 1$ provided that $B(t) \ge 1$. Hence on the set $\{B(t) > 0\}$ we can define the sequences $\mu = \mu(t)$ and $\nu_2 = \nu_2(t) \le \nu_3 = \nu_3(t) \le \ldots \le \nu_\mu = \nu_\mu(t) \le t$ as follows:

$$\nu_k = \inf\{s : 0 \le s \le t, \ Q(s,t) \ge k\}$$

and $\mu = B(t)$ is the largest positive integer for which $\nu_\mu \le t$.

Now we prove our

LEMMA 7.6 *For any k fixed we have*

$$\mathbf{E}((t-\nu_k) \mid B(t) \geq k) \sim \frac{t}{k}$$

and

$$\mathbf{E}((t-\nu_k)^2 \mid B(t) \geq k) \sim \frac{2t^2}{k(k+1)}.$$

Proof. On the set $\{k \leq \mu\} = \{B(t) \geq k\}$ by (7.2) we have

$$\mathbf{E}(B(t) \mid \nu_k,\ B(t) \geq k) \sim \frac{k(t-\nu_k)\sigma^2}{2}. \tag{7.18}$$

Hence

$$\frac{1}{\mathbf{P}\{k \leq \mu\}} \int_{\{k\leq\mu\}} \mathbf{E}(B(t) \mid \nu_k) d\mathbf{P} \sim \frac{k\sigma^2}{2\mathbf{P}\{k \leq \mu\}} \left(t\mathbf{P}\{k \leq \mu\} - \int_{\{k\leq\mu\}} \nu_k d\mathbf{P} \right) = $$
$$= \frac{k\sigma^2}{2}(t - \mathbf{E}(\nu_k \mid k \leq \mu)).$$

Since

$$\frac{1}{\mathbf{P}\{k \leq \mu\}} \int_{\{k\leq\mu\}} \mathbf{E}(B(t) \mid \nu_k) d\mathbf{P} = \mathbf{E}(B(t) \mid k \leq \mu) =$$
$$= \mathbf{E}(B(t) \mid B(t) \geq k) \sim \mathbf{E}(B(t) \mid B(t) > 0) \sim \frac{t\sigma^2}{2},$$

we have the first statement of Lemma 7.6.

Observe that

$$\mathrm{Var}\,(B(t) \mid B(t) > 0) \sim \frac{\sigma^4 t^2}{4}$$

and

$$\mathrm{Var}\,(B(t) \mid \nu_k,\ B(t) \geq k) \sim \frac{k\sigma^4}{4}(t-\nu_k)^2. \tag{7.19}$$

Then by (7.18) and (7.19)

$$\mathbf{E}\mathrm{Var}\,(B(t) \mid \nu_k,\ B(t) \geq k) =$$
$$= \mathbf{E}(B^2(t) \mid B(t) \geq k) - \mathbf{E}((\mathbf{E}(B(t) \mid \nu_k,\ B(t) \geq k))^2) \sim$$
$$\sim \mathbf{E}(B^2(t) \mid B(t) > 0) - \mathbf{E}((\mathbf{E}(B(t) \mid \nu_k,\ B(t) \geq k))^2) \sim$$
$$\sim \frac{\sigma^2 t}{2}(\sigma^2 t + 1) - \mathbf{E}\left(\frac{k^2\sigma^4}{4}(t-\nu_k)^2 \;\middle|\; B(t) \geq k \right) \sim \mathbf{E}\left(\frac{k\sigma^4}{4}(t-\nu_k)^2 \;\middle|\; B(t) \geq k \right).$$

Hence

$$\mathbf{E}((t-\nu_k)^2 \mid B(t) \geq k) \sim \frac{2t(\sigma^2 t + 1)}{k(k+1)\sigma^2} \sim \frac{2t^2}{k(k+1)}$$

and we have the second statement of Lemma 7.6.

We can also evaluate the limit distribution of ν_k.

THEOREM 7.2 *For any k fixed and $0 \le x \le 1$ we have*

$$\lim_{t\to\infty} \mathbf{P}\left\{\frac{\nu_k}{t} < x \,\middle|\, B(t) \ge k\right\} = x^{k-1}.$$

Proof. Let $B_1(t), B_2(t), \ldots, B_k(t)$ be i.i.d.r.v.'s with

$$B_i(t) \stackrel{\mathcal{D}}{=} B(t) \qquad (i = 1, 2, \ldots, k)$$

and let

$$B_1(t) + B_2(t) + \cdots + B_k(t) = \Gamma(t).$$

Then by (7.3) we have

$$\mathbf{P}\left\{\frac{\Gamma(t)}{t} = y \,\middle|\, B_i(t) > 0,\ i = 1, 2, \ldots, k\right\} = \frac{\lambda^k y^{k-1}}{(k-1)!} e^{-\lambda y} \quad (y \ge 0)$$

where

$$\lambda = \frac{2}{\sigma^2}.$$

Let

$$f_k(x) = \mathbf{P}\left\{\frac{\nu_k}{t} = x \,\middle|\, B(t) \ge k\right\}.$$

Then

$$\begin{aligned}
&\mathbf{P}\left\{\frac{B(t)}{t} = z \,\middle|\, \nu_k = tx,\ B(t) \ge k\right\} = \\
&= \mathbf{P}\left\{\frac{B_1(t(1-x)) + \cdots + B_k(t(1-x))}{t} = z \,\middle|\, \nu_k = tx,\ B(t) \ge k\right\} = \\
&= \frac{\lambda^k}{(k-1)!}\left(\frac{z}{1-x}\right)^{k-1} \exp\left(-\lambda\frac{z}{1-x}\right)\frac{1}{1-x}.
\end{aligned}$$

Hence

$$\begin{aligned}
\lambda e^{-\lambda z} &\sim \mathbf{P}\left\{\frac{B(t)}{t} = z \,\middle|\, B(t) > 0\right\} \sim \mathbf{P}\left\{\frac{B(t)}{t} = z \,\middle|\, B(t) \ge k\right\} \sim \\
&\sim \int_0^1 \frac{\lambda^k}{(k-1)!}\left(\frac{z}{1-x}\right)^{k-1} \exp\left(-\lambda\frac{z}{1-x}\right)\frac{1}{1-x} f_k(x)\,dx.
\end{aligned}$$

It is easy to see that the only solution of the integral–equation

$$\lambda e^{-\lambda z} = \int_0^1 \frac{\lambda^k}{(k-1)!}\left(\frac{z}{1-x}\right)^{k-1} \exp\left(-\lambda\frac{z}{1-x}\right)\frac{1}{1-x} f_k(x)\,dx$$

is

$$f_k(x) = (k-1)x^{k-2} \qquad (0 \le x \le 1).$$

This, in turn, implies Theorem 7.2.

Consider a fixed $[0,T]$-branch $\{S_k\}_{k=0}^T$ of the underlying branching random walk. Clearly $\{S_k\}_{k=0}^T$ is a random walk. Let $\xi_1 = \xi_1(T) = \nu_2$ be the first time-point where from a new (ξ_1, T)-branch is starting which lives up to time T. That is to say the particle located in S_{ξ_1} at time ξ_1 has at least two children living up to time T. The first one follows the branch $S_{x_1+1}, S_{\xi_1+2}, \ldots, S_T$. The second one executes an independent branching random walk starting from S_{ξ_1} at time ξ_1.

Let $\xi_2, \xi_3, \ldots$ be the second, third, ... such points. That is to say the particle located in S_{ξ_2} at time ξ_2 has at least two children living up to time T. The first one follows the branch $S_{\xi_2+1}, S_{\xi_2+2}, \ldots, S_T$. The second one executes an independent branching random walk starting from S_{ξ_2} at time ξ_2.

Now consider a partition of the $B(T) - 1$ particles living at time T, not considering the terminal point of the fixed branch $\{S_k\}_{k=0}^T$. The first class $C_1 = C_1(T)$ consists of the terminal points of those (ξ_1, T)-branches which are branching from the fixed branch $\{S_k\}_{k=0}^T$ at ξ_1. $C_2 = C_2(T)$ consists of the terminal points of those (ξ_2, T)-branches which are branching from the fixed branch $\{S_k\}_{k=0}^T$ at ξ_2. See Fig. 2.

Let $U_1, U_2, \ldots$ be a sequence of independent, $[0,1]$-uniform r.v.'s and introduce the following notations:

$$V_1 = 1, \qquad V_k = \prod_{j=1}^{k-1}(1-U_j) \quad (k = 2,3,\ldots),$$

$$L_k = \sum_{j=1}^{k} U_j V_j \quad (k = 1,2,\ldots).$$

Then we have

LEMMA 7.7 *For any k fixed*

$$\lim_{T\to\infty} \mathbf{P}\left\{ \frac{\xi_1}{T} < x_1, \frac{\xi_2}{T} < x_2, \ldots, \frac{\xi_k}{T} < x_k \,\middle|\, B(T) > 0 \right\} =$$

$$\mathbf{P}\{L_1 < x_1, L_2 < x_2, \ldots, L_k < x_k\}$$

and

$$\lim_{T\to\infty} (T - \xi_k)^{-1} \mathbf{E}(|C_k| \mid \xi_k,\ B(T) > 0) = \frac{\sigma^2}{2}.$$

Proof. Our second statement is a trivial consequence of (7.2). In case $k = 1$ our first statement follows from Theorem 7.2. In case $k = 2$ observe that

$$\frac{\xi_2}{T} = \frac{\xi_2 - \xi_1}{T - \xi_1}\left(1 - \frac{\xi_1}{T}\right) + \frac{\xi_1}{T}.$$

Clearly

$$\lim_{T\to\infty} \mathbf{P}\left\{ \frac{\xi_2 - \xi_1}{T - \xi_1} < x \,\middle|\, \xi_1,\ B(T) > 0 \right\} = \mathbf{P}\{U_2 < x\}.$$

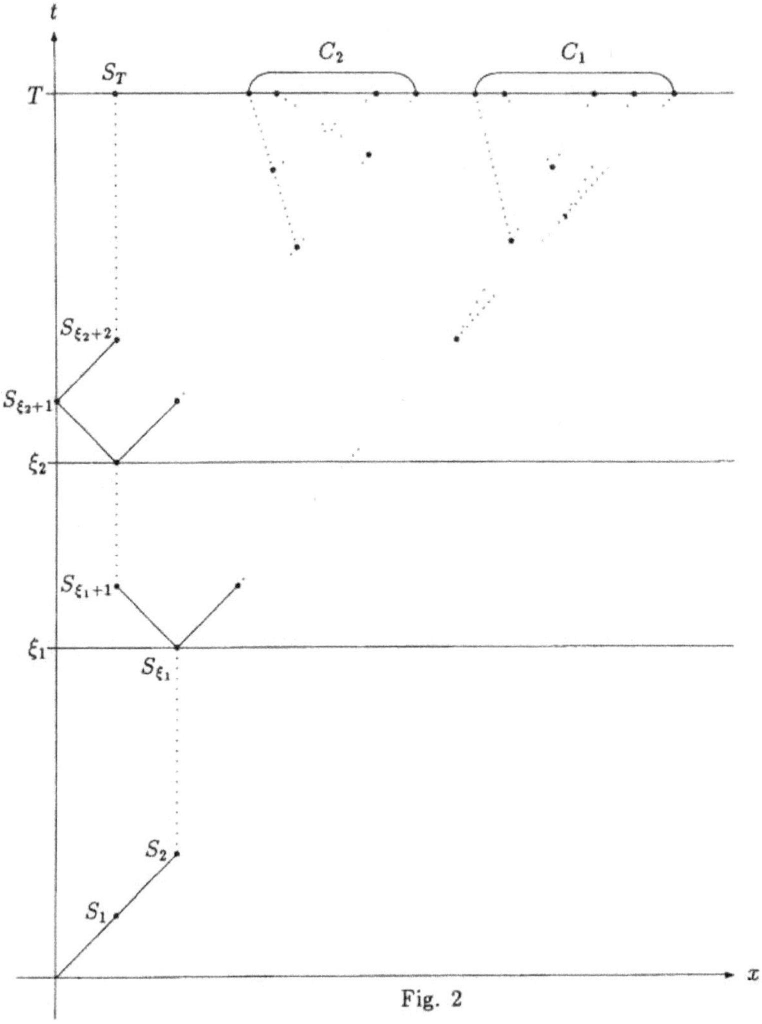

Fig. 2

Hence

$$\lim_{T\to\infty} \mathbf{P}\left\{ \frac{\xi_2}{T} < x_2, \frac{\xi_1}{T} < x_1 \,\middle|\, B(T) > 0 \right\} = \mathbf{P}\{U_2(1-U_1) + U_1 < x_2, U_1 < x_1\}$$

i.e. we have our first statement in case $k = 2$.

Observing that

$$\frac{\xi_3}{T} = \frac{\xi_3 - \xi_2}{T - \xi_2}\left(1 - \frac{\xi_2}{T}\right) + \frac{\xi_2}{T},$$

$$\lim_{T\to\infty} \mathbf{P}\left\{ \frac{\xi_3 - \xi_2}{T - \xi_2} < x \,\middle|\, \xi_2,\ B(T) > 0 \right\} = \mathbf{P}\{U_3 < x\}$$

$$\lim_{T\to\infty} \mathbf{P}\left\{ \frac{\xi_2}{T} < x \,\middle|\, B(T) > 0 \right\} = \mathbf{P}\{U_2(1-U_1) + U_1 < x\}$$

we have Lemma 7.7 in case $k = 3$.

Continuing this procedure we complete our proof by induction.

7.2 On the expectation of $\lambda(x,t)$

The model and the notations of Sections 4.2 and 4.3 (with $m = 1$) will be used without any further discussion. For sake of simplicity in this Section we agree that $0/0 = 0$.

LEMMA 7.8

$$\mathbf{E}(Z(x,t,\mu) \mid B(t), B(t+1)) = \frac{B(t+1)}{B(t)} \tag{7.20}$$

$t = 0, 1, 2, \ldots,\ x \in \mathbb{Z}^d,\ \mu = 1, 2, \ldots, \lambda(x,t)$.

Proof. Since

$$\sum_{x\in\mathbb{Z}^d} \sum_{\mu=1}^{\lambda(x,t)} Z(x,t,\mu) = B(t+1),$$

we have

$$B(t+1) = \mathbf{E}(B(t+1) \mid B(t), B(t+1)) = \sum_{x\in\mathbb{Z}^d} \sum_{\mu=1}^{\lambda(x,t)} \mathbf{E}(Z(x,t,\mu) \mid B(t), B(t+1)) =$$
$$= B(t)\mathbf{E}(Z(x,t,\mu) \mid B(t), B(t+1))$$

and (7.20) is proved.

LEMMA 7.9 (cf. Lemma 4.3) *Let*

$$g(x,T,t) = \frac{B(T)}{B(t)} f(x,T,t) = \frac{B(T)}{B(t)} \sum_{y \in \mathbb{Z}^d} \lambda(y,t) p(y \leadsto x, T-t).$$

Then

$$\mathbf{E}\left(\frac{\lambda(x,T)}{B(T)} \;\middle|\; \mathcal{F}(t), B(T) \right) = \frac{g(x,T,t)}{B(T)}. \tag{7.21}$$

Proof. Since

$$\mathbf{E}\left(\frac{\lambda(x,T)}{B(T)} \;\middle|\; \mathcal{F}(T), B(T) \right) = \frac{\lambda(x,T)}{B(T)}$$

and

$$g(x,T,T) = f(x,T,T) = \sum_{y \in \mathbb{Z}^d} \lambda(y,T) p(y \leadsto x, 0) = \lambda(x,T),$$

we have (7.21) in case $t = T$. In case $T = 1,\ t = 0$ we have

$$\mathbf{E}\left(\frac{\lambda(x,1)}{B(1)} \;\middle|\; \mathcal{F}(0), B(1) \right) = \begin{cases} \frac{1}{2d} & \text{if } \ x \in \mathcal{C}(0,1), \\ 0 & \text{if } \ x \notin \mathcal{C}(0,1) \end{cases}$$

and

$$\frac{g(x,1,0)}{B(1)} = \sum_{y \in \mathbb{Z}^d} \lambda(y,0) p(y \leadsto x, 1) = p(0 \leadsto x, 1).$$

Consequently we have (7.21) in case $T = 1,\ t = 0$ as well. Now we give the proof by induction. Assume that (7.21) holds true for T, for any $0 \le t \le T$ and $x \in \mathbb{Z}^d$. Then

$$\mathbf{E}\left(\frac{\lambda(x,T+1)}{B(T+1)} \;\middle|\; \mathcal{F}(t), B(T+1) \right) =$$
$$= \mathbf{E}\left(\mathbf{E}\left(\frac{\lambda(x,T+1)}{B(T+1)} \;\middle|\; \mathcal{F}(T), B(T+1) \right) \;\middle|\; \mathcal{F}(t), B(T+1) \right)$$

and by Lemma 7.8

$$\mathbf{E}(\lambda(x,T+1) \mid \mathcal{F}(T), B(T+1)) =$$
$$= \mathbf{E}\left(\sum_{y \in \mathcal{C}(x)} \sum_{\mu=1}^{\lambda(y,T)} I_{x-y}(X(y,T,\mu)) Z(y,T,\mu) \;\middle|\; \mathcal{F}(T), B(T+1) \right) =$$
$$= \frac{1}{2d} \frac{B(T+1)}{B(T)} \sum_{y \in \mathcal{C}(x)} \lambda(y,T).$$

Hence by the condition of the induction

$$\mathbf{E}\left(\frac{\lambda(x,T+1)}{B(T+1)}\;\middle|\;\mathcal{F}(t),B(T+1)\right)=\frac{1}{2d}\mathbf{E}\left(\frac{1}{B(T)}\sum_{y\in C(x)}\lambda(y,T)\;\middle|\;\mathcal{F}(t),B(T+1)\right)=$$
$$=\frac{1}{2d}\mathbf{E}\left(\mathbf{E}\left(\frac{1}{B(T)}\sum_{y\in C(x)}\lambda(y,T)\;\middle|\;\mathcal{F}(t),B(T),B(T+1)\right)\;\middle|\;\mathcal{F}(t),B(T+1)\right)=$$
$$=\frac{1}{2d}\mathbf{E}\left(\mathbf{E}\left(\frac{1}{B(T)}\sum_{y\in C(x)}\lambda(y,T)\;\middle|\;\mathcal{F}(t),B(T)\right)\;\middle|\;\mathcal{F}(t),B(T+1)\right)=$$
$$=\frac{1}{2d}\mathbf{E}\left(\sum_{y\in C(x)}\frac{g(y,T,t)}{B(T)}\;\middle|\;\mathcal{F}(t),B(T+1)\right).$$

Note that

$$\frac{1}{2d}\sum_{y\in C(x)}\frac{g(y,T,t)}{B(T)}=\frac{1}{B(t)}\frac{1}{2d}\sum_{y\in C(x)}\sum_{u\in\mathbb{Z}^d}\lambda(u,t)p(u\rightsquigarrow y,T-t)=$$
$$=\frac{1}{B(t)}\sum_{u\in\mathbb{Z}^d}\lambda(u,t)p(u\rightsquigarrow x,T+1-t)=\frac{g(x,T+1,t)}{B(T+1)}$$

is measurable with respect to $\mathcal{F}(t)\cup B(T+1)$, which implies Lemma 7.9.

LEMMA 7.10 (cf. Lemma 4.5)

$$\mathbf{E}(\lambda(x,T)\mid B(T))=B(T)p(0\rightsquigarrow x,T), \tag{7.22}$$
$$\mathbf{E}(g(x,T,t)\mid \mathcal{F}(t-1),B(T))=g(x,T,t-1), \tag{7.23}$$
$$\mathbf{E}(g(x,T,t)\mid B(T))=B(T)p(0\rightsquigarrow x,T). \tag{7.24}$$

Proof. By Lemma 7.9

$$\mathbf{E}\left(\frac{\lambda(x,T)}{B(T)}\;\middle|\;B(T)\right)=\mathbf{E}\left(\frac{\lambda(x,T)}{B(T)}\;\middle|\;\mathcal{F}(0),B(T)\right)=\frac{g(x,T,0)}{B(T)}=p(0\rightsquigarrow x,T)$$

and we have (7.22). Similarly

$$\mathbf{E}(g(x,T,t)\mid\mathcal{F}(t-1),B(T))=\mathbf{E}(\mathbf{E}(\lambda(x,T)\mid\mathcal{F}(t),B(T))\mid\mathcal{F}(t-1),B(T))=$$
$$=\mathbf{E}(\lambda(x,T)\mid\mathcal{F}(t-1),B(T))=g(x,T,t-1)$$

and we have (7.23). Since by (7.21) and (7.22)

$$\mathbf{E}(g(x,T,t)\mid B(T))=\mathbf{E}(\mathbf{E}(\lambda(x,T)\mid\mathcal{F}(t),B(T))\mid B(T))=$$
$$=\mathbf{E}(\lambda(x,T)\mid B(T))=B(T)p(0\rightsquigarrow x,T)$$

which implies (7.24).

7.3 What might we expect?

Comparing (7.22) and (4.6) we see that the expectations $\mathbf{E}(\lambda(x,t)(B(t))^{-1})$ on the set $\{B(t)>0\}$ in cases $m=1$ and $m>1$ behave very similarly. Hence we might expect that the limit properties of $\lambda(x,t)/B(t)$ (as $t\to\infty$) in cases $m=1$ and $m>1$ are also similar. An important difference between the two cases that given $\{B(t)>0\}$ in cases $m=1$ resp. $m>1$ $B(t)$ is growing like t resp. m^t. Since $|\mathcal{C}(0,t^{1/2})|=O(t^{d/2})$, only in case $d=1$ we might hope that the t particles will be distributed normally. It turns out that even in case $d=1$ the distribution of the particles is very strange. This statement is a consequence of the following:

THEOREM 7.3 *There exists a* $0<\varepsilon<1$ *such that*

$$\varepsilon\le\mathbf{P}\{\lambda(0,2t)=0\mid B(2t)>0\}\le 1-\varepsilon \tag{7.25}$$

for any $t=1,2,\dots$ *and* $d=1$. *Further*

$$\lim_{t\to\infty}\mathbf{P}\{\lambda(0,t)=0\mid B(t)>0\}=1 \tag{7.26}$$

if $d\ge 3$.

Instead of proving the upper part of (7.25) we prove the following much stronger

THEOREM 7.4 *For any* $d=1,2,\dots,$ $\delta>0,$ $K>0,$ $\|x\|\le Kt^{1/2}$ *there exists an* $\varepsilon=\varepsilon(\delta,K)>0$ *such that*

$$\mathbf{P}\left\{\sum_{z\notin\mathcal{C}(x,\delta t^{1/2})}\lambda(z,t)=0\mid B(t)>0\right\}\ge\varepsilon.$$

Proof. At time ν_k $(k=2,3,\dots,\mu)$ there are $k-1$ particles which live up to t. Let the locations of these particles at time ν_k be $X_{k1},X_{k2},\dots,X_{k,k-1}$.

Clearly for some $0<\varepsilon_0<1$ and $0<c_1<c_2<1$

$$\mathbf{P}\{c_1t<\nu_2<c_2t\}\ge\varepsilon_0$$

and

$$\mathbf{P}\left\{\|X_{21}-x\|\le\frac{\delta}{2}t^{1/2}\right\}\ge\varepsilon_0.$$

Similarly for any $\lambda_1>0$ we have

$$\mathbf{P}\{\max_{1\le i\le 3}\|X_{4i}-X_{21}\|\ge\lambda_1(\nu_4-\nu_2)^{1/2}\}\le 3C\exp\left(-\frac{\lambda_1^2}{2}\right)$$

where $C>0$ is an absolute constant. In general

$$\mathbf{P}\{A_\ell\mid\nu_{2^\ell},\nu_{2^{\ell+1}}\}\le 2^{2\ell+1}C\exp\left(-\frac{\lambda_\ell^2}{2}\right)$$

where

$$A_\ell = \left\{ \max_{\substack{1\le i\le 2^\ell-1 \\ j\in \mathcal{A}(i,\ell)}} \|X_{2^{\ell+1},j} - X_{2^\ell,i}\| \ge \lambda_\ell(\nu_{2^{\ell+1}} - \nu_{2^\ell})^{1/2} \right\}$$

and $j \in \mathcal{A}(i,\ell)$ if and only if $X_{2^{\ell+1},j}$ is a descendant of $X_{2^\ell,i}$. Hence we have

$$\mathbf{P}\left\{\bigcap_{\ell=1}^{\kappa} \overline{A}_\ell \mid \nu_2, \nu_4, \ldots, \nu_{2^{\kappa+1}}\right\} \ge \prod_{\ell=1}^{\kappa} \left(1 - 2^{2\ell+1} C \exp\left(-\frac{\lambda_\ell^2}{2}\right)\right)$$

where $2^{\kappa+1} = t$. Observe that by Lemma 7.6

$$\mathbf{E}((\nu_{2^\ell} - \nu_{2^{\ell-1}})^{1/2} \mid B(t) > 0) \le (\mathbf{E}(\nu_{2^\ell} - \nu_{2^{\ell-1}} \mid B(t) > 0))^{1/2} \le$$
$$\le \left(\frac{2^\ell - 1}{2^\ell} - \frac{2^{\ell-1} - 1}{2^{\ell-1}}\right)^{1/2} t^{1/2} = \left(\frac{t}{2^{\ell-1}}\right)^{1/2}.$$

Hence one can choose the sequence $\{\lambda_\ell\}$ such that

$$\prod_{\ell=1}^{\kappa} \left(1 - 2^{\ell+1} C \exp\left(-\frac{\lambda_\ell^2}{2}\right)\right) \ge \varepsilon$$

and

$$\sum_{\ell=1}^{\kappa} \lambda_\ell (\nu_{2^{\ell+1}} - \nu_{2^\ell})^{1/2} \le \varepsilon t^{1/2}$$

with positive probability and Theorem 7.4 is proved.

The upper part of (7.25) is a trivial consequence of Theorem 7.4.

Instead of proving the lower part of (7.25) we prove the following much stronger

THEOREM 7.5 *There exists a $0 < \varepsilon < 1$ such that*

$$\mathbf{P}\{\lambda(0, 2t) \ge \varepsilon t^{1/2} \mid B(2t) > 0\} \ge \varepsilon$$

for any $t = 1, 2, \ldots$ and $d = 1$.

Before the proof we give a few lemmas.

LEMMA 7.11 *Let $\{S_1(k)\}_{k=0}^\infty$, $\{S_2(k)\}_{k=0}^\infty$ and $\{S_3(k)\}_{k=0}^\infty$ three independent random walks. Define the non-independent random walks $T_1(\cdot)$ and $T_2(\cdot)$ as follows:*

$$T_1(k) = T_1(k,s) = \begin{cases} S_1(k) & \text{if } k \le s, \\ S_1(s) + S_2(k-s) & \text{if } k > s, \end{cases}$$

$$T_2(k) = T_2(k,s) = \begin{cases} S_1(k) & \text{if } k \le s, \\ S_1(k) + S_3(k-s) & \text{if } k > s \end{cases}$$

where $s = 1, 2, \ldots$. Further let

$$I_1 = I_1(n) = \begin{cases} 1 & \text{if} \quad T_1(n) = 0, \\ 0 & \text{if} \quad T_1(n) \neq 0, \end{cases}$$

$$I_2 = I_2(n) = \begin{cases} 1 & \text{if} \quad T_2(n) = 0, \\ 0 & \text{if} \quad T_2(n) \neq 0 \end{cases}$$

where $n = s + 1, s + 2, \ldots$.

Then

$$\mathbf{E} I_1 I_2 \sim (2\pi)^{-1}(n^2 - s^2)^{-1/2}$$

if $s \to \infty$ and $n - s \to \infty$.

Proof. Clearly we have

$$\begin{aligned}
&\mathbf{E} I_1 I_2 = \mathbf{P}\{I_1 = I_2 = 1\} = \\
&= \sum_{k=-\infty}^{+\infty} \mathbf{P}\{I_1 = I_2 = 1 \mid S_1(s) = k\} \mathbf{P}\{S_1(s) = k\} = \\
&= \sum_{k=-\infty}^{+\infty} \mathbf{P}\{S_2(n-s) = S_3(n-s) = -k\} \mathbf{P}\{S_1(s) = k\} \sim \\
&\sim \sum_{k=-\infty}^{+\infty} \left((2\pi(n-s))^{-1/2} \exp\left(-\frac{k^2}{2(n-s)} \right) \right)^2 (2\pi s)^{-1/2} \exp\left(-\frac{k^2}{2s} \right) \sim \\
&\sim \frac{1}{(2\pi)^{3/2}} \frac{1}{s^{1/2}(n-s)} \int_{-\infty}^{+\infty} \exp\left(-\frac{x^2}{2} \left(\frac{2}{n-s} + \frac{1}{s} \right) \right) dx = \frac{1}{2\pi(n^2 - s^2)^{1/2}}.
\end{aligned}$$

Hence Lemma 7.11 is proved.

LEMMA 7.12 *Consider the terminal point S_T of a fixed $[0, T]$-branch $\{S_k\}_{k=0}^{T}$ of the underlying branching random walk. Let R_k be an arbitrary element of the class C_k (cf. Lemma 7.7). Further let*

$$I_0 = \begin{cases} 1 & \text{if} \quad S_T = 0, \\ 0 & \text{if} \quad S_T \neq 0, \end{cases}$$

$$I_k = \begin{cases} 1 & \text{if} \quad R_k = 0, \\ 0 & \text{if} \quad R_k \neq 0. \end{cases}$$

Then for any k fixed

$$\mathbf{P}\{I_0 = I_k = 1 \mid \xi_k,\ B(T) > 0\} \sim (2\pi)^{-1}(T^2 - \xi_k^2)^{-1/2}$$

as $T \to \infty$.

Proof. It is a trivial consequence of Lemma 7.11.

LEMMA 7.13

$$\mathbf{E}(\lambda(0,2T) \mid B(2T) > 0) \sim \pi^{-1/2}\sigma^2 T^{1/2}$$

as $T \to \infty$ and for any $\varepsilon > 0$

$$\mathrm{Var}\,(\lambda(0,2T) \mid B(2T) > 0) \le (1+\varepsilon)\frac{\sigma^4 T}{\pi}\frac{3-2^{3/2}}{2^{3/2}-1}$$

if T is big enough.

Proof. Our first statement is a simple consequence of (7.22).

Let $P_1, P_2, \ldots, P_{B(2T)}$ be the locations of the particles at time $2T$. Then

$$\lambda(0,2T) = \sum_{i=1}^{B(2T)} I(P_i)$$

where

$$I(x) = \begin{cases} 1 & \text{if} \quad x = 0, \\ 0 & \text{if} \quad x \neq 0. \end{cases}$$

Hence

$$\begin{aligned} &\mathrm{Var}\,(\lambda(0,2T) \mid B(2T) > 0) = \\ &= 2\mathbf{E}\left(\Big(\sum_{1 \le i < j \le B(2T)} I(P_i)I(P_j) \;\Big|\; B(2T) > 0 \right) + \mathbf{E}(\lambda(0,2T) \mid B(2T) > 0) - \\ &\quad - (\mathbf{E}(\lambda(0,2T) \mid B(2T) > 0))^2 \sim \\ &\sim 2\mathbf{E}\left(\sum_{1 \le i < j \le B(2T)} I(P_i)I(P_j) \;\Big|\; B(2T) > 0 \right) + \pi^{-1/2}\sigma^2 T^{1/2} - \pi^{-1}\sigma^4 T. \end{aligned}$$

Consider a fixed $[0,2T]$-branch $\{S_k\}_{k=0}^{2T}$ of the underlying branching random walk. For sake of simplicity we assume that $S_{2T} = P_1$. Let $P_j \in C_k$ (cf. Lemma 7.7). Then by Lemma 7.12

$$\mathbf{E}(I(P_1)I(P_j) \mid \xi_k,\; B(2T) > 0) \sim (2\pi)^{-1}(4T^2 - \xi_k^2)^{-1/2}.$$

By Lemma 7.7 we have

$$\mathbf{E}\left(\sum_{P_j \in C_k} I(P_1)I(P_j) \;\Big|\; \xi_k,\; B(2T) > 0 \right) \sim \frac{\sigma^2}{4\pi}\left(\frac{1-(2T)^{-1}\xi_k}{1+(2T)^{-1}\xi_k}\right)^{1/2}$$

and

$$2\mathbf{E}\left(\sum_{1 \le i < j \le B(2T)} I(P_i)I(P_j) \;\Big|\; B(2T) > 0 \right) \sim$$

$$
\begin{aligned}
&\sim T\frac{\sigma^4}{2\pi}\sum_{k=1}^{\infty}\mathbf{E}\left(\frac{1-(2T)^{-1}\xi_k}{1+(2T)^{-1}\xi_k}\right)^{1/2} \le \\
&\le T\frac{\sigma^4}{2\pi}\sum_{k=1}^{\infty}(\mathbf{E}(1-(2T)^{-1}\xi_k))^{1/2} \sim T\frac{\sigma^4}{2\pi}\sum_{k=1}^{\infty}(\mathbf{E}(1-L_k))^{1/2} = \\
&= T\frac{\sigma^4}{2\pi}\sum_{k=1}^{\infty}\left(1-\sum_{j=1}^{k}2^{-j}\right)^{1/2} = T\frac{\sigma^4}{2\pi}\sum_{k=1}^{\infty}2^{-k/2} = \frac{\sigma^4 T}{2\pi(2^{1/2}-1)}
\end{aligned}
$$

which, in turn, implies Lemma 7.13.

Proof of Theorem 7.5. By Lemma 7.13

$$\mathrm{Var}\,(\lambda(0,2T)\mid B(2T)>0) \le (\mathbf{E}(\lambda(0,2T)\mid B(2T)>0))^2.$$

Hence we have Theorem 7.5 by Chebyshev inequality.

Now, we turn to the proof of (7.26). In fact we prove the following stronger

THEOREM 7.6 *Let $d \ge 3$. Then*

$$O(t^{-d/2}) \le \mathbf{P}\{\lambda(0,t)\neq 0 \mid B(t)>0\} \le O(t^{-(d-2)/3}). \tag{7.27}$$

Assume that $m_\mu < \infty$. Then

$$O(t^{-d/2}) \le \mathbf{P}\{\lambda(0,t)\neq 0 \mid B(t)>0\} \le O(t^{-\mu(d-2)/2(1+\mu)}). \tag{7.28}$$

Proof. Let $1<\alpha<d/2$. Then by (7.13) we have

$$
\begin{aligned}
&\mathbf{P}\{\lambda(0,t)\neq 0 \mid B(t)>0\} = \\
&= \mathbf{P}\{\lambda(0,t)\neq 0 \mid 0<B(t)<t^\alpha\}\mathbf{P}\{B(t)<t^\alpha \mid B(t)>0\} + \\
&\quad + \mathbf{P}\{\lambda(0,t)\neq 0 \mid B(t)>t^\alpha\}\mathbf{P}\{B(t)>t^\alpha \mid B(t)>0\} \le \\
&\le \mathbf{P}\{\lambda(0,t)\neq 0 \mid 0<B(t)<t^\alpha\} + \mathbf{P}\{B(t)>t^\alpha \mid B(t)>0\} \le \\
&\le O\left(\frac{t^\alpha}{t^{d/2}}\right) + O\left(\frac{1}{t^{2\alpha-2}}\right).
\end{aligned}
$$

Choosing $\alpha = (d+4)/6$ we have the upper part of (7.27). Its lower part is a trivial consequence of Lemma 4.1.

The proof of (7.28) is the same as that of (7.27).

Note that, assuming $m_\mu<\infty$, for any $x \in C(0,KT^{1/2})$ we have

$$O(t^{-d/2}) \le \mathbf{P}\{\lambda(x,t)\neq 0 \mid B(t)>0\} \le O(t^{-\mu(d-2)/2(1+\mu)}). \tag{7.29}$$

Theorem 7.3 does not say anything about the case $d=2$. Very likely in this case (7.26) holds true.

We note that Theorem 7.4 in some sense gives the best possible result. In fact we have

THEOREM 7.7 *Let $\delta_t (t = 1, 2, \dots)$ be a sequence of positive numbers with $\delta_t \searrow 0$. Then*

$$\lim_{t\to\infty} \mathbf{P}\left\{ \bigcup_{x\in\mathbb{Z}^d} \left\{ \sum_{z\notin C(x,\delta_t,t^{1/2})} \lambda(z,t) = 0 \right\} \;\middle|\; B(t) > 0 \right\} = 0.$$

Proof is trivial.

Theorem 7.4 told us that it can happen (with positive probability) that all particles living at time t are concentrated in a small ball. Our next theorem claims that, in contrary, it can happen (with positive probability) that the particles living at time t are spread over in a large ball. In fact we have

THEOREM 7.8 *For any $d = 1, 2, \dots$, $K > 0$ there exists an $\varepsilon = \varepsilon(K) > 0$ such that with probability $\geq \varepsilon$ one can find K particles living at time t, located in $P_1(t), P_2(t), \dots, P_K(t)$ such that*

$$\inf_{1\leq i<j\leq K} \|P_i(t) - P_j(t)\| \geq K t^{1/2}.$$

Proof is trivial.

Let

$$a(t)\#\{x : \; x \in \mathbb{Z}^d, \; \lambda(x,t) > 0\}$$

be the number of occupied locations at time t. In case $d = 1$ Theorem 7.4 implies that for any $\delta > 0$ there exists an $\varepsilon = \varepsilon(\delta) > 0$ such that

$$\mathbf{P}\{a(t) < \delta t^{1/2}\} \geq \varepsilon$$

i.e. $a(t)$ can be smaller than $\delta t^{1/2}$.

Let δ_t be a sequence of positive numbers with $\delta_t \to \infty$. Then it is easy to see that

$$\lim_{t\to\infty} \mathbf{P}\{a(t) \leq \delta_t t^{1/2} \mid B(t) > 0\} = 1.$$

The next theorem tells us that $a(t)$ cannot be smaller than $t^{1/3}$.

THEOREM 7.9 *For any $\delta > 0$ we have*

$$\lim_{t\to\infty} \mathbf{P}\{a(t) \geq t^{1/3-\delta} \mid B(t) > 0\} = 1.$$

Proof. Let $0 < \alpha < 1$ and consider the $B(t - t^\alpha)$ particles living at time $t - t^\alpha$. The probability that a particle living at time $t - t^\alpha$ still lives at time t is $O(t^{-\alpha})$. Note that for any $\varepsilon > 0$ there exists a $\delta_1 = \delta_1(\varepsilon) > 0$ such that

$$\lim_{t\to\infty} \mathbf{P}\{B(t - t^\alpha) \geq \delta_1 t \mid B(t) > 0\} \geq 1 - \varepsilon.$$

Hence with probability $1-2\varepsilon$ among the $B(t-t^{\alpha})$ particles living at time $t-t^{\alpha}$, $\delta_1 t^{1-\alpha}/2$ particles will live at time t. Since these $\delta_1 t^{1-\alpha}/2$ particles move independently, they will occupy at time t at least $t^{\alpha/2}$ different locations provided that $t^{1-\alpha} \gg t^{\alpha/2}$ i.e. $\alpha < 2/3$. Hence we have Theorem 7.9.

Very likely in Theorem 7.9 the lower bound $t^{1/3-\delta}$ is very weak. We guess that it can be replaced by $t^{1/2-\delta}$.

Finally we propose an unsolved problem in case $d = 1$. By Theorem 7.5 $\lambda(0,2T) \geq \varepsilon T^{1/2}$ with positive probability provided that $B(2T) > 0$. It looks very likely that there is a long interval $(-a,b)$ around the origin such that $\lambda(x,2T) > 0$ if $x \in (-a,b)$ and $\lambda(0,2T) \geq \varepsilon T^{1/2}$. More precisely let

$$\begin{aligned} A(T) &= \sup\{x: \ x<0, \ \lambda(x,2T)=0\}, \\ B(T) &= \inf\{x: \ x>0, \ \lambda(x,2T)=0\}, \\ C(T) &= A(T)+B(T). \end{aligned}$$

Then our problem is to characterize the class of those functions $f(T)$ for which $C(T)/f(T)$ has a conditional limit distribution given $\lambda(0,2T) \geq \varepsilon T^{1/2}$. It is easy to see that for any $\delta > 0$

$$\lim_{T\to\infty} \mathbf{P}\{C(T) \geq T^{1/4-\delta} \mid \lambda(0,2T) \geq \varepsilon T^{1/2}\} = 1.$$

A natural conjecture is: the exponent $1/4-\delta$ can be replaced by $1/2-\delta$.

It looks also interesting in case $d=1$ to study the limit properties of $\max_{x\in\mathbb{Z}^1}\lambda(x,t)$ given $B(t) > 0$. Here we formulate only the following trivial

THEOREM 7.10 *Let $d = 1$. Then*

(i) *there exists an $\varepsilon > 0$ such that*

$$\mathbf{P}\{\max_{x\in\mathbb{Z}^1} \lambda(x,t) \geq \varepsilon t^{1/2} \mid B(t) > 0\} \geq 1-\varepsilon$$

for any $t = 1,2,\ldots$,

(ii) *for any $K > 0$ there exists an $\varepsilon = \varepsilon(K) > 0$ such that*

$$\mathbf{P}\{\max_{x\in\mathbb{Z}^1} \lambda(x,t) \geq K t^{1/2} \mid B(t) > 0\} \geq \varepsilon$$

for any $t = 1,2,\ldots$.

Chapter 8

Critical branching random walks of a random field

8.1 The number of visits in the origin

Here we use the notations and conditions of Section 5.1, however we assume that

$$\sum_{i=0}^{\infty} ip_i = m = 1.$$

Let

$$I = \#\{t:\ t = 0, 1, 2, \ldots,\ \Lambda(0,t) > 0\}$$

and

$$I(x) = \#\{t:\ t = 0, 1, 2, \ldots,\ \lambda_x(0,t) > 0\}$$

i.e. I is the total number of those time points when the location 0 is occupied and $I(x)$ is the number of those time points when the location 0 is occupied by one of the offsprings of the particle located originally in x.

THEOREM 8.1

$$\mathbf{P}\{I = \infty\} = \begin{cases} 0 & \text{if} \quad d = 1, \\ 1 & \text{if} \quad d \geq 2. \end{cases}$$

In case $d = 1$ we prove a much stronger theorem. In fact we prove that there exists a big empty interval around the origin if t is big enough.

THEOREM 8.2 *Let $d = 1$ and*

$$c = c(t) = \max\left\{x:\ x \geq 0,\ \sum_{y=0}^{x} \Lambda(y,t) = 0\right\}$$

and

$$\xi = \xi(t) = \min\{x:\ x \geq 0,\ B_x(t) > 0\}.$$

Then we have

$$\lim_{t\to\infty} \mathbf{P}\{t^{-1}\xi(t) > x\} = \exp\left(-\frac{2x}{\sigma^2}\right) \qquad (x > 0), \tag{8.1}$$

$$\lim_{t\to\infty} \mathbf{P}\{t^{-1}c(t) > x\} = \exp\left(-\frac{2x}{\sigma^2}\right) \qquad (x > 0). \tag{8.2}$$

Further for any $\varepsilon > 0$ we have

$$\lim_{t\to\infty} \frac{\xi(t)}{t(\log t)^{1+\varepsilon}} = 0 \qquad a.s., \tag{8.3}$$

$$\lim_{t\to\infty} \frac{c(t)}{t(\log t)^{1+\varepsilon}} = 0 \qquad a.s., \tag{8.4}$$

$$\lim_{t\to\infty} \frac{(\log t)^{1+\varepsilon}}{t}\xi(t) = \infty \quad a.s., \tag{8.5}$$

$$\lim_{t\to\infty} \frac{(\log t)^{1+\varepsilon}}{t}c(t) = \infty \quad a.s. \tag{8.6}$$

Proof. By (7.1) we have

$$\mathbf{P}\{\xi = k\} = \left(1 - \frac{2 + o_t(1)}{\sigma^2 t}\right)^k \frac{2 + o_t(1)}{\sigma^2 t} \qquad (k = 0, 1, 2, \ldots),$$

$$\mathbf{P}\{\xi \geq k\} = \left(1 - \frac{2 + o_t(1)}{\sigma^2 t}\right)^k \qquad (k = 0, 1, 2, \ldots)$$

and

$$\lim_{t\to\infty} \mathbf{P}\{t^{-1}\xi(t) > x\} = \exp\left(-\frac{2}{\sigma^2}x\right) \qquad (x > 0).$$

Hence we have (8.1).

We easily obtain also that for any $\varepsilon > 0$

$$\mathbf{P}\{\xi(t) > t(\log t)^{1+\varepsilon}\} \sim \exp\left(-\frac{2}{\sigma^2}(\log t)^{1+\varepsilon}\right),$$

$$\mathbf{P}\left\{\xi(t) < \frac{t}{(\log t)^{1+\varepsilon}}\right\} \sim \frac{2}{\sigma^2(\log t)^{1+\varepsilon}} \qquad (\varepsilon > 0).$$

Hence we have (8.3). In order to get (8.5) let $t_k = [e^k]$ $(k = 1, 2, \ldots)$. Then

$$\xi(t_k) \geq t_k(\log t_k)^{-1-\varepsilon} \qquad \text{a.s.}$$

for all but finitely many k. Let $t_k \leq \tau < t_{k+1}$. Then

$$\xi(\tau) \geq \xi(t_k) \geq t_k(\log t_k)^{-1-\varepsilon} \geq \tau(\log \tau)^{-1-2\varepsilon}$$

if k is big enough. Hence we have

$$\xi(t) \geq \frac{t}{(\log t)^{1+\varepsilon}} \qquad \text{a.s.}$$

for all but finitely many t and we have (8.5).

Observe that for any $\varepsilon > 0$ we have

$$\xi(t) - t^{1/2+\varepsilon} \le c(t) \le \xi(t) + t^{1/2+\varepsilon} \qquad \text{a.s.} \tag{8.7}$$

for all but finitely many t. It follows from the fact that any particle (or any element of a set of particles consisting not more than t particles) cannot move farther from its starting point than $t^{1/2+\varepsilon}$.

Then (8.2), (8.4) and (8.6) follow from (8.1), (8.3), (8.5) and (8.7).

In case $d = 2$ instead of Theorem 8.1 we prove the following stronger

THEOREM 8.3 *Let $d = 2$ and*

$$Z_k = \#\{x : \ x \in \mathbb{Z}^2, \ n_k \le |x| < n_{k+1}, \ I(x) > 0\}$$

where

$$n_k = [\exp(\exp(Ck \log k))] \qquad (C > 0).$$

Then

$$Z_k \ge 1 \qquad a.s.$$

for all but finitely many k if C is big enough.

Proof. Let

$$\tau(u, v) = \#\{x : \ x \in \mathbb{Z}^2, \ |x| = u, \ B_x(v) > 0\}.$$

Then for any $\alpha > 0$ by (7.1) we have

$$\mathbf{P}\{\tau([\alpha r], r^2) \ge 1\} \ge O(r^{-1}) \qquad (d = 2) \tag{8.8}$$

as $r \to \infty$. We recall the following:

LEMMA 8.1 ([44] Theorem 19.3) *Let $d = 2$ and $S(\cdot)$ be a simple, symmetric random walk on $\mathbb{Z}^2$. Then for any $x \in \mathbb{Z}^2$ with $|x| = [\alpha r]$ we have*

$$\mathbf{P}\{\exists k : \ 0 \le k \le r^2, \ S(k) = x\} = O\left(\frac{1}{\log r}\right)$$

provided that $0 < \alpha \le 1/20$.

By Lemma 8.1 and (8.8) we have

$$\mathbf{P}\{\exists x : \ |x| = [\alpha r], \ I(x) > 0\} = O\left(\frac{1}{r \log r}\right).$$

Hence

$$\mathbf{P}\{Z_k = 0\} = \prod_{r=n_k}^{n_{k+1}-1} \left(1 - \frac{O(1)}{r \log r}\right) \le \exp\left(-\sum_{r=n_k}^{n_{k+1}-1} \frac{O(1)}{r \log r}\right) =$$

$$= \exp\left(-O(1) \log \frac{\log(n_{k+1} - 1)}{\log n_k}\right) = \exp(-O(1) C \log k).$$

Consequently

$$\sum_{k=1}^{\infty} \mathbf{P}\{Z_k = 0\} < \infty$$

if C big enough, which, in turn, implies Theorem 8.3.

Similarly, in case $d \geq 3$ instead of Theorem 8.1 we prove the following stronger

THEOREM 8.4 *Let $d \geq 3$ and define Z_k just like in Theorem 8.3 with*

$$n_k = [\exp(Ck \log k)] \qquad (C > 0).$$

Then

$$Z_k \geq 1 \qquad a.s.$$

for all but finitely many k if C is big enough.

Proof. Let $d \geq 3$ and apply (7.1). Then for any $K > 0$

$$\mathbf{E}\tau(r, r^2 K) = O(r^{d-3}). \tag{8.9}$$

Now we recall the following:

LEMMA 8.2 ([44] Lemma 22.15) *For any $x \in \mathbb{Z}^d$ $(d \geq 3)$ let*

$$J(x) = \begin{cases} 0 & \text{if} \quad S(n) \neq x \text{ for all } n = 0, 1, 2, \ldots, \\ 1 & \text{otherwise.} \end{cases}$$

Then

$$\mathbf{P}\{J(x) = 1\} = \frac{C_d + o(1)}{|x|^{d-2}}.$$

Let

$$J(x, N) = \begin{cases} 0 & \text{if} \quad S(n) \neq x \text{ for all } n = 0, 1, 2, \ldots, N, \\ 1 & \text{otherwise.} \end{cases}$$

Then we prove

LEMMA 8.3 *Let $|x| = [\alpha r]$, $d \geq 3$. Then for any $\varepsilon > 0$ there exists a $K = K(\varepsilon) > 0$ such that*

$$\mathbf{P}\{J(x, |x|^2 K) = 1\} \geq (C_d - \varepsilon)|x|^{2-d}.$$

Proof. Clearly

$$\mathbf{P}\{J(x) = 1, J(x, |x|^2 K) = 0\} \leq \sum_{k=Kr^2}^{\infty} \mathbf{P}\{S(k) = x\} < 3\left(\frac{d}{2\pi}\right)^{d/2} \sum_{k=Kr^2}^{\infty} k^{-d/2} \leq$$

$$\leq \frac{8}{d-2}\left(\frac{d}{2\pi}\right)^{d/2} K^{1-d/2} r^{2-d} \leq \varepsilon r^{2-d}$$

if K is big enough.

Since

$$\{J(x)=1\}=\{J(x,|x|^2K)=1\}\cup\{J(x)=1,J(x,|x|^2K)=0\},$$

we have Lemma 8.3.

By (8.9) and Lemma 8.3 we have

$$\mathbf{P}\{\exists x:\ |x|=\lfloor\alpha r\rfloor,\ I(x)>0\}\geq O\left(\frac{r^{d-3}}{r^{d-2}}\right)=O(r^{-1}).$$

Hence

$$\mathbf{P}\{Z_k=0\}\leq\prod_{r=n_k}^{n_{k+1}-1}\left(1-\frac{O(1)}{r}\right)=k^{-O(1)C}$$

and we have Theorem 8.4.

8.2 Clusters in case $d=1$

Our recent goal is to prove that in case $d=1$ those integers x's for which $\Lambda(x,t)>0$ are concentrated in small clusters.

Let

$$\begin{aligned}
A(\xi,t)&=\{y:\ y\in\mathbb{Z}^1,\ 0\leq y\leq\xi,\ \Lambda(y,t)>0\},\\
A(\xi,x,t)&=\{y:\ y\in\mathbb{Z}^1,\ 0\leq y\leq\xi,\ \lambda_x(y,t)>0\},\\
A(t)&=A(t,t),\qquad A_x(t)=A(t,x,t),\\
a(\xi,t)&=|A(\xi,t)|,\qquad a(\xi,x,t)=|A(\xi,x,t)|,\\
a(t)&=a(t,t),\qquad a_x(t)=a(t,x,t).
\end{aligned}$$

LEMMA 8.4 *For any $K>0$, $x\in\mathbb{Z}^1$ and $\varepsilon>0$ we have*

$$\mathbf{P}\{d(A_x(t),x)\geq t^{1/2+\varepsilon}\}\leq Ct^{-K}\tag{8.10}$$

where $C=C(\varepsilon,K)>0$ and

$$d(A_x(t),x)=\begin{cases}\max\limits_{z\in A_x(t)}|z-x| & \text{if}\quad A_x(t)\neq\emptyset,\\ 0 & \text{if}\quad A_x(t)=\emptyset.\end{cases}$$

Consequently

$$\mathbf{P}\{a_x(t)\geq t^{1/2+\varepsilon}\}\leq Ct^{-K}.$$

Proof. It is a trivial consequence of (7.13).

THEOREM 8.5 *For any* $\varepsilon > 0$

$$a(t) \leq t^{1/2+\varepsilon} \qquad a.s. \tag{8.11}$$

for all but finitely many t.

Proof. Let

$$\mathcal{B}(t) = \{x : \ x \in \mathbb{Z}^1, \ x \in [-t, 2t], \ B_x(t) > 0\}$$

and

$$b(t) = |\mathcal{B}(t)|.$$

Then by (7.1) for any $\rho > 0$

$$\mathbf{P}\{b(t) \geq \rho \log t\} \leq O(t^{-\rho}). \tag{8.12}$$

(8.11) clearly follows from (8.10) and (8.12).

Now we wish to find a lower estimate of $a(t)$.

THEOREM 8.6 *For any* $\varepsilon > 0$ *there exists a* $p = p(\varepsilon) > 0$ *such that*

$$\mathbf{P}\{a(t) \geq t^{1/3-\varepsilon}\} \geq p(\varepsilon) \qquad (t = 0, 1, 2, \ldots). \tag{8.13}$$

Further there exists a $\vartheta > 0$ *such that*

$$\mathbf{P}\{a(t) = 0\} \geq \vartheta \qquad (t = 0, 1, 2, \ldots). \tag{8.14}$$

Consequently

$$a(t) = 0 \qquad i.o. \quad a.s.$$

and

$$\limsup_{t\to\infty} \frac{a(t)}{t^{1/3-\varepsilon}} = \infty \qquad a.s.$$

Proof. By (7.1) we have

$$\mathbf{P}\{a(t) = 0\} \geq \mathbf{P}\left\{\sum_{x=-t}^{2t} B_x(t) = 0\right\} =$$
$$= (\mathbf{P}\{B_0(t) = 0\})^{3t} \to \exp\left(-\frac{6}{\sigma^2}\right) \qquad \text{as} \qquad t \to \infty.$$

Hence we have (8.14).

Let $0 < \alpha < 1$ and

$$c(t) = \sum_{y=0}^{t} \Lambda(y, t - t^{\alpha}).$$

Then by (7.1) and (7.3) there exists a positive constant $p_1 = p_1(\alpha) > 0$ such that

$$\mathbf{P}\{c(t) > t\} \geq p_1.$$

Any particle living at time $t - t^\alpha$ will live at t with probability $O(t^{-\alpha})$. Hence with positive probability $t^{1-\alpha}$ particles among the $c(t)$ particles will still live at t. Since these $t^{1-\alpha}$ particles move independently, they will occupy at time t at least $t^{\alpha/2}$ different locations provided that $t^{1-\alpha} \gg t^{\alpha/2}$ i.e. $\alpha < 2/3$. Hence we have (8.13).

It looks also interesting to study the geometrical properties of the set $A(t)$. As we have seen (cf. (8.12)) the number of particles originated in $[-t, 2t]$ and still living at time t cannot be more that $2\log t$. None of these particles can produce more than t^3 offsprings (living at t) (cf. (7.5)). Clearly (cf. Lemma 8.4) the distance between the ancestor and any of its offsprings is not more that $t^{1/2+\varepsilon}$. Hence we have

THEOREM 8.7 *For any $\varepsilon > 0$ there exists a random sequence*

$$0 \leq x_1 < x_2 < \ldots < x_{[2\log t]} \leq t$$

of integers such that

$$A(t) \subset \bigcup_{i=1}^{[2\log t]} [x_i - t^{1/2+\varepsilon}, x_i + t^{1/2+\varepsilon}] \qquad a.s.$$

for all but finitely many t.

This theorem clearly means that at time t in $[0, t]$ one can find $[2\log t]$ clusters of size $t^{1/2+\varepsilon}$ where the particles living at time t and located in $[0, t]$ (i.e. the set $A(t)$) are concentrated.

We have seen that $a(t) = a(t, t) = 0$ with positive probability.

Now, we note that for any $\varepsilon > 0$ we have

$$\lim_{t\to\infty} a(t^{1+\varepsilon}, t) = \infty \qquad \text{a.s.}$$

In fact

THEOREM 8.8 *For any $\varepsilon > 0$*

$$\lim_{t\to\infty} t^{-1/2+\varepsilon/2} a(t^{1+\varepsilon}, t) = \infty \qquad a.s.$$

Proof follows easily by the above given methods.

It looks also interesting to study the limit properties of

$$L(t) = \sum_{y=1}^{t} \lambda(y, t).$$

By (7.1) and (7.2)

$$\mathbf{E}L(t) = t \qquad (t = 0, 1, 2, \ldots).$$

The following theorem gives more information

THEOREM 8.9 *We have*

$$\lim_{t\to\infty} \mathbf{P}\{L(t) < xt\} = G(x) \qquad (x \geq 0)$$

where

$$G(x) = \mathbf{P}\left\{\sum_{i=1}^{\pi} E_i < x\right\},$$

$\pi, E_1, E_2, \ldots$ *are independent r.v.'s with*

$$\mathbf{P}\{\pi = k\} = \frac{1}{k!}\left(\frac{2}{\sigma^2}\right)^k \exp\left(-\frac{2}{\sigma^2}\right) \qquad (k = 0, 1, 2, \ldots),$$

$$\mathbf{P}\{E_i < x\} = 1 - \exp\left(-\frac{2x}{\sigma^2}\right) \qquad (i = 1, 2, \ldots;\ x \geq 0).$$

Proof. Here we present only the idea of the proof. Let

$$\mathcal{B}^*(t) = \{x :\ x \in \mathbb{Z}^1,\ x \in [0, t],\ B_x(t) > 0\},$$
$$b^*(t) = |\mathcal{B}^*(t)|.$$

Then by (7.1)

$$\lim_{t\to\infty} \mathbf{P}\{b^*(t) = k\} = \lim_{t\to\infty} \binom{t}{k} \left(\frac{2}{t\sigma^2}\right)^k \left(1 - \frac{2}{t\sigma^2}\right)^{t-k} =$$
$$= \frac{1}{k!}\left(\frac{2}{\sigma^2}\right)^k \exp\left(-\frac{2}{\sigma^2}\right) \qquad (k = 0, 1, 2, \ldots)$$

i.e. the number of particles located originally in $[0, t]$ and living up to time t is a Poisson r.v. in limit as $t \to \infty$. Then applying (7.3) we have the Theorem. The details can be obtained by the methods of proofs of Theorems 8.4 and 8.5.

Remark 8.1. Assume that the set $\mathcal{B}^*(t)$ is known and we consider a particle living at time t, located in $[0, t]$ and we ask about its ancestor. Then we can give exactly its ancestor in $\mathcal{B}^*(t)$ almost surely for all but finitely many t. This fact can be seen by the methods given above.

8.3 No clusters in case $d \geq 2$

In this Section we do not give exact theorems, rather we intend to give an intuitive description of the shape of the located points.

Introduce the following notations:

(i) for any $(i_1, i_2, \ldots, i_d) \in \mathbb{Z}^d$ let

$$A = A(i_1, i_2, \ldots, i_d, t) = [i_1, i_1 + t^{1/d}) \times [i_2, i_2 + t^{1/d}) \times \cdots \times [i_d, i_d + t^{1/d}),$$

(ii) $b = b(t) = b(i_1, i_2, \ldots, i_d, t) = \#\{x: \; x \in \mathbb{Z}^d A, \; B_x(t) > 0\}$,

(iii) $\mathcal{B} = \mathcal{B}(t) = \{x: \; x \in \mathbb{Z}^d, \; B_x(t) > 0\}$,

(iv) for each $x \in \mathcal{B}$ define a random walk $\{S_x(k), \; k = 0, 1, \ldots, t\}$ with $S_x(0) = x$ and assume that these random walks are independent,

(v) $\mathcal{D} = \mathcal{D}(t) = \{S_x(t) : \; x \in \mathcal{B}\}$,

(vi) $d = d(t) = \#\{S_x(t) : \; S_x(t) \in \mathbb{Z}^d A, \; x \in \mathcal{B}\}$.

LEMMA 8.5

$$\mathbf{P}\{x \in \mathcal{B}\} = \frac{2}{\sigma^2 t}(1 + o(1)) \qquad (x \in \mathbb{Z}^d),$$

$$\mathbf{P}\{y \in \mathcal{D}\} = \frac{2}{\sigma^2 t}(1 + o(1)) \qquad (y \in \mathbb{Z}^d),$$

$$\lim_{t \to \infty} \mathbf{P}\{b(t) = k\} = \frac{1}{k!}\left(\frac{2}{\sigma^2}\right)^k \exp\left(-\frac{2}{\sigma^2}\right) \quad (k = 0, 1, 2, \ldots),$$

$$\lim_{t \to \infty} \mathbf{P}\{d(t) = k\} = \frac{1}{k!}\left(\frac{2}{\sigma^2}\right)^k \exp\left(-\frac{2}{\sigma^2}\right) \quad (k = 0, 1, 2, \ldots).$$

Proof is trivial.

Let $x \in \mathcal{B}(t)$. Then at time t, for any $\varepsilon > 0$ there are $B_x(t)$ living particles located in $\mathcal{C}(S_x(t), t^{1/2+\varepsilon})$. Hence at time t in any cube $A(i_1, i_2, \ldots, i_d, t)$ one can find finitely many (in fact Poisson $2/\sigma^2$) points such that the balls of radius $t^{1/2+\varepsilon}$ around these points consist of $O(t)$ particles. This means that

in case $d = 1$ in $[0, t)$ there are not more than $[2 \log t]$ clusters of radius $t^{1/2+\varepsilon}$ (cf. Theorem 8.7),

in case $d = 2$ a cube A may contain the offsprings of a single particle only, i.e. the particles located in A at time t can be brothers, hence it can happen that clusters (non-well separated) can be found,

in case $d \geq 3$ the offsprings of a single particle will never be concentrated in a single cube A, hence there are no clusters.

Chapter 9

Multitype branching random walk

9.1 Multitype branching process

Throughout this Chapter for sake of simplicity we consider only two–type processes.

At time $t = 0$ we have two particles of different types, type 0 and type 1. At time $t = 1$ the particle of type i ($i = 0, 1$) produces k ($k = 0, 1, 2, \ldots$) offsprings of type 0 and ℓ ($\ell = 0, 1, 2, \ldots$) offsprings of type 1 with probability $p(i, k, \ell)$. Producing these offsprings the particle dies. In case $k = \ell = 0$ we say that the process dies out. In case $k + \ell > 0$ each offspring repeats this procedure independently.

More formally speaking: let $Y(i, \mu, j, t)$ ($i = 0, 1$; $\mu = 0, 1, 2, \ldots$; $j = 0, 1$; $t = 1, 2, \ldots$) be an array of independent r.v.'s with

$$\mathbf{P}\{Y(i, \mu, 0, t) = k, \; Y(i, \mu, 1, t) = \ell\} = p(i, k, \ell)$$

where

$$p(i, k, \ell) \geq 0 \qquad (i = 0, 1; \; k = 0, 1, 2, \ldots; \; \ell = 0, 1, 2, \ldots)$$

and

$$\sum_{k=0}^{\infty} \sum_{\ell=0}^{\infty} p(i, k, \ell) = 1 \qquad (i = 0, 1).$$

Define the array $\{B(i, t), \; i = 0, 1; \; t = 0, 1, 2, \ldots\}$ as follows:

$$\begin{aligned}
B(0,0) &= B(1,0) = 1, \\
B(0,1) &= Y(0,1,0,1) + Y(1,1,0,1), \\
B(1,1) &= Y(0,1,1,1) + Y(1,1,1,1), \\
B(0,2) &= \sum_{\mu=1}^{B(0,1)} Y(0,\mu,0,2) + \sum_{\mu=1}^{B(1,1)} Y(1,\mu,0,2), \\
B(1,2) &= \sum_{\mu=1}^{B(0,1)} Y(0,\mu,1,2) + \sum_{\mu=1}^{B(1,1)} Y(1,\mu,1,2), \\
\ldots & \qquad \ldots\ldots \\
B(i,t) &= \sum_{\mu=1}^{B(0,t-1)} Y(0,\mu,i,t) + \sum_{\mu=1}^{B(1,t-1)} Y(1,\mu,i,t).
\end{aligned}$$

Here $Y(i, \mu, j, t)$ can be considered as the number of those offsprings of type j which are produced by the μ–th particle of type i at time t. $B(i, t)$ ($i = 0, 1$; $t = 0, 1, 2, \ldots$) is the number of type i particles at time t.

From now on we assume that

$$\mathbf{E}Y(i,\mu,0,t)=\sum_{k=0}^{\infty}\sum_{\ell=0}^{\infty}kp(i,k,\ell)=m_{i0}<\infty \qquad (i=0,1),$$
$$\mathbf{E}Y(i,\mu,1,t)=\sum_{k=0}^{\infty}\sum_{\ell=0}^{\infty}\ell p(i,k,\ell)=m_{i1}<\infty \qquad (i=0,1)$$

and

$$0<\operatorname{Var}Y(i,\mu,0,t)=\sum_{k=0}^{\infty}\sum_{\ell=0}^{\infty}(k-m_{i0})^2p(i,k,\ell)=\sigma_{i0}^2<\infty \qquad (i=0,1),$$
$$0<\operatorname{Var}Y(i,\mu,1,t)=\sum_{k=0}^{\infty}\sum_{\ell=0}^{\infty}(\ell-m_{i1})^2p(i,k,\ell)=\sigma_{i1}^2<\infty \qquad (i=0,1).$$

We are interested in the limit properties of the two-type branching process $\{B(i,t),\ i=0,1\}$ as $t\to\infty$.

Let

$$M=\begin{pmatrix} m_{00} & m_{01}\\ m_{10} & m_{11}\end{pmatrix},\qquad M^t=\begin{pmatrix} m_{00}^{(t)} & m_{01}^{(t)}\\ m_{10}^{(t)} & m_{11}^{(t)}\end{pmatrix}$$
$$(t=1,2,\ldots;\ m_{ij}^{(1)}=m_{ij}),$$
$$B(t)=(B(0,t),B(1,t)) \qquad (t=0,1,2,\ldots),$$
$$\mathcal{B}(t)=\mathcal{B}(B(s),\ 0\le s\le t)$$

be the smallest σ-algebra with respect to which the random vectors $B(s)$ $(0\le s\le t)$ are measurable.

Note that

$$\mathbf{E}B(0,t)=m_{00}^{(t)}+m_{10}^{(t)} \qquad (t=1,2,\ldots), \tag{9.1}$$
$$\mathbf{E}B(1,t)=m_{01}^{(t)}+m_{11}^{(t)} \qquad (t=1,2,\ldots), \tag{9.2}$$
$$\mathbf{E}(B(0,t+1)\mid\mathcal{B}(t))=m_{00}B(0,t)+m_{10}B(1,t) \qquad (t=0,1,2,\ldots) \tag{9.3}$$

and

$$\mathbf{E}(B(1,t+1)\mid\mathcal{B}(t))=m_{01}B(0,t)+m_{11}B(1,t) \qquad (t=0,1,2,\ldots). \tag{9.4}$$

(9.3) and (9.4) combined tell us

$$\mathbf{E}(B(t+1)\mid\mathcal{B}(t))=B(t)M. \tag{9.5}$$

In case $t=1$ (9.1), (9.2), (9.3) and (9.4) are trivial. In general, they can be obtained by induction.

We will see that the limit properties of the sequence $\{B(t),\ t = 0,1,2,\ldots\}$ strongly depend on the structure of the matrix M. At first as a characteristic we consider the number of zeros in M. Let

$$I(x) = \begin{cases} 1 & \text{if} \quad x = 0, \\ 0 & \text{if} \quad x \neq 0, \end{cases}$$

$$I(M) = \sum_{i=0}^{1}\sum_{j=0}^{1} I(m_{ij}).$$

We note that

(a) in case $I(M) = 0$, $\det M = 0$ there exists a $\mu > 0$ such that $m_{10} = \mu m_{00}$ and $m_{11} = \mu m_{01}$,

(b) in case $I(M) = 0$, $\det M \neq 0$, M has two different, real eigenvalues λ_0, λ_1 and two linearly independent, real eigenvectors

$$x_0 = \begin{pmatrix} x_{00} \\ x_{01} \end{pmatrix}, \qquad x_1 = \begin{pmatrix} x_{10} \\ x_{11} \end{pmatrix}$$

such that

$$Mx_i = \lambda_i x_i \qquad (i = 0,1) \tag{9.6}$$

and

$$\lambda_0 = \frac{1}{2}(S - (S^2 - 4D)^{1/2}) =$$
$$= \frac{1}{2}(m_{00} + m_{11} - ((m_{00} - m_{11})^2 + 4m_{01}m_{10})^{1/2}) \neq 0 \tag{9.7}$$
$$\lambda_1 = \frac{1}{2}(S + (S^2 - 4D)^{1/2}) =$$
$$= \frac{1}{2}(m_{00} + m_{11} + ((m_{00} - m_{11})^2 + 4m_{01}m_{10})^{1/2}), \tag{9.8}$$
$$\lambda_1 > |\lambda_0| > 0 \tag{9.9}$$

where

$$S = m_{00} + m_{11}, \quad D = \det M = m_{00}m_{11} - m_{01}m_{10}.$$

Now we present two simple lemmas for later reference.

LEMMA 9.1 *Assume that $I(M) = 0$, $\det M \neq 0$. Then we have*

$$\lambda_1 > \max(m_{00}, m_{11}), \qquad \lambda_0 < \min(m_{00}, m_{11}), \tag{9.10}$$
$$x_{10} > 0, \qquad x_{11} > 0, \tag{9.11}$$
$$x_{00} > 0, \qquad x_{01} < 0, \tag{9.12}$$
$$\det \Xi > 0 \tag{9.13}$$

where

$$\Xi = \begin{pmatrix} x_{00} & x_{10} \\ x_{01} & x_{11} \end{pmatrix}.$$

Proof. (9.6) implies that

$$x_0 = \begin{pmatrix} 1 \\ (\lambda_0 - m_{00})/m_{01} \end{pmatrix} \quad \text{and} \quad x_1 = \begin{pmatrix} 1 \\ (\lambda_1 - m_{00})/m_{01} \end{pmatrix}$$

are eigenvectors. Hence (9.9) implies (9.10), (9.11) and (9.12). (9.13) follows from (9.11) and (9.12).

The following lemma is wellknown in linear algebra.

LEMMA 9.2 *Assume that* $I(M) = 0$ *and* $\det M \neq 0$. *Then*

$$M = \Xi \Lambda \Xi^{-1}$$

where

$$\Lambda = \begin{pmatrix} \lambda_0 & 0 \\ 0 & \lambda_1 \end{pmatrix}$$

and

$$M^t = \begin{pmatrix} m_{00}^{(t)} & m_{01}^{(t)} \\ m_{10}^{(t)} & m_{11}^{(t)} \end{pmatrix} = \Xi \Lambda^t \Xi^{-1} \quad (t = 1, 2, \ldots)$$

where

$$\Lambda^t = \begin{pmatrix} \lambda_0^t & 0 \\ 0 & \lambda_1^t \end{pmatrix}.$$

Let

$$\Xi^{-1} = \begin{pmatrix} y_{00} & y_{10} \\ y_{01} & y_{11} \end{pmatrix}.$$

Then assuming the conditions of Lemma 9.2 we have

$$\begin{aligned} m_{00}^{(t)} &= \lambda_0^t x_{00} y_{00} + \lambda_1^t x_{10} y_{01} = (\lambda_1^t - \lambda_0^t) x_{10} y_{01} + \lambda_0^t, \\ m_{01}^{(t)} &= \lambda_0^t x_{00} y_{10} + \lambda_1^t x_{10} y_{11} = (\lambda_1^t - \lambda_0^t) x_{10} y_{11}, \\ m_{10}^{(t)} &= \lambda_0^t x_{01} y_{00} + \lambda_1^t x_{11} y_{01} = (\lambda_1^t - \lambda_0^t) x_{11} y_{01}, \\ m_{11}^{(t)} &= \lambda_0^t x_{01} y_{10} + \lambda_1^t x_{11} y_{11} = (\lambda_1^t - \lambda_0^t) x_{11} y_{11} + \lambda_0^t. \end{aligned} \tag{9.14}$$

Still assuming the conditions of Lemma 9.2 we have

$$\mathbf{E}(B(t+1)x_i \mid \mathcal{B}(t)) = B(t) M x_i = \lambda_i B(t) x_i \qquad (i = 0, 1)$$

or

$$\mathbf{E}(B(t+1)\Xi \mid \mathcal{B}(t)) = \Lambda B(t) \Xi.$$

Hence

$$\mathcal{M}_0(t) = \frac{B(0,t)x_{00} + B(1,t)x_{01}}{\lambda_0^t}$$

and

$$\mathcal{M}_1(t) = \frac{B(0,t)x_{10} + B(1,t)x_{11}}{\lambda_1^t}$$

are martingales.

By (9.1), (9.2), (9.6) and (9.11) we have

$$\mathbf{E}\mathcal{M}_1(t) = \mathbf{E}|\mathcal{M}_1(t)| = (m_{00}^{(t)} + m_{10}^{(t)})x_{10} + (m_{01}^{(t)} + m_{11}^{(t)})x_{11} = \lambda_1^t(x_{10} + x_{11})$$

and

$$\mathbf{E}\mathcal{M}_0(t) = (m_{00}^{(t)} + m_{10}^{(t)})x_{00} + (m_{01}^{(t)} + m_{11}^{(t)})x_{01} = \lambda_0^t(x_{00} + x_{01}).$$

Then by the martingale convergence theorem ([24] p. 319, Th. 4.1) there exists a r.v. $B \geq 0$ with $\mathbf{E}B = 1$ such that

$$\lim_{t\to\infty} \mathcal{M}_1(t) = (x_{10} + x_{11})B. \qquad \text{a.s.} \tag{9.15}$$

Now we formulate our first result.

THEOREM 9.1 (i) *If* $I(M) = 0$, $\det M = 0$ *and* $m_{00} + \mu m_{01} = m_{00} + m_{11} \leq 1$ *then*

$$\lim_{t\to\infty} B(0,t) = \lim_{t\to\infty} B(1,t) = 0 \qquad a.s.$$

(ii) *If* $I(M) = 0$, $\det M = 0$ *and* $m_{00} + \mu m_{01} > 1$ *then there exists a r.v.* $B \geq 0$ *with* $\mathbf{E}B = 1$ *such that*

$$\lim_{t\to\infty} \frac{B(0,t)}{(m_{00} + \mu m_{01})^t} = m_{00}(1+\mu)B \qquad a.s.$$

and

$$\lim_{t\to\infty} \frac{B(1,t)}{(m_{00} + \mu m_{01})^t} = m_{01}(1+\mu)B \qquad a.s.$$

(iii) *If* $I(M) = 0$, $\det M \neq 0$ *and* $\lambda_1 \leq 1$ *then*

$$\lim_{t\to\infty} B(0,t) = \lim_{t\to\infty} B(1,t) = 0 \qquad a.s.$$

(iv) *If* $I(M) = 0$, $\det M \neq 0$ *and* $\lambda_1 > 1$ *then there exists a r.v.* $B \geq 0$ *with* $\mathbf{E}B = 1$ *such that*

$$\lim_{t\to\infty} \frac{B(0,t)}{\lambda_1^t} = -\frac{(x_{10} + x_{11})x_{01}}{\det \Xi} B \quad a.s.$$

and

$$\lim_{t\to\infty} \frac{B(1,t)}{\lambda_1^t} = \frac{(x_{10} + x_{11})x_{00}}{\det \Xi} B \quad a.s.$$

Proof of (i) and (ii). Observe that by (9.3) and (9.4) we have

$$\mathbf{E}(B(0,t+1)\mid \mathcal{B}(t)) = m_{00}(B(0,t)+\mu B(1,t)) \qquad (t=0,1,2,\ldots) \tag{9.16}$$

and

$$\mathbf{E}(B(1,t+1)\mid \mathcal{B}(t)) = m_{01}(B(0,t)+\mu B(1,t)) \qquad (t=0,1,2,\ldots). \tag{9.17}$$

Hence

$$\mathbf{E}(B(0,t+1)+\mu B(1,t+1)\mid \mathcal{B}(t)) = (m_{00}+\mu m_{01})(B(0,t)+\mu B(1,t)). \tag{9.18}$$

By (9.17)

$$\mathcal{M}(t) = \frac{B(0,t)+\mu B(1,t)}{(m_{00}+\mu m_{01})^t}$$

is a martingale. Hence by the martingale convergence theorem there exists a r.v. $B \geq 0$ with $\mathbf{E}B = 1$ such that

$$\lim_{t\to\infty} \mathcal{M}(t) = (1+\mu)B \qquad \text{a.s.}$$

This implies (i).

In case (ii) by (9.16), (9.17) and the law of large numbers on the set $\{B>0\}$ we have

$$\lim_{t\to\infty} \frac{B(0,t+1)}{B(0,t)+\mu B(1,t)} = m_{00} \qquad \text{a.s.}$$

and

$$\lim_{t\to\infty} \frac{B(1,t+1)}{B(0,t)+\mu B(1,t)} = m_{01} \qquad \text{a.s.}$$

In turn, we have (ii).

Proof of (iii). By (9.15) we have

$$\lim_{t\to\infty} (x_{10}B(0,t)+x_{11}B(1,t)) = 0 \qquad \text{a.s.}$$

which, in turn, implies (iii).

Proof of (iv). By the law of large numbers for any fixed integer $\nu \geq 1$ we have

$$\lim_{t\to\infty} \frac{B(1,t+\nu)}{m_{01}^{(\nu)}B(0,t)+m_{11}^{(\nu)}B(1,t)} = 1 \quad \text{a.s.} \tag{9.19}$$

and

$$\lim_{t\to\infty} \frac{B(0,t+\nu)}{m_{00}^{(\nu)}B(0,t)+m_{10}^{(\nu)}B(1,t)} = 1 \quad \text{a.s.} \tag{9.20}$$

Consider

$$\frac{m_{01}^{(\nu)}B(0,t)+m_{11}^{(\nu)}B(1,t)}{\lambda_1^{t+\nu}}-\frac{x_{10}B(0,t)+x_{11}B(1,t)}{\lambda_1^t}y_{11}=$$

$$=\frac{B(0,t)}{\lambda_1^t}\frac{m_{01}^{(\nu)}-x_{10}y_{11}\lambda_1^\nu}{\lambda_1^\nu}+\frac{B(1,t)}{\lambda_1^t}\frac{m_{11}^{(\nu)}-x_{11}y_{11}\lambda_1^\nu}{\lambda_1^\nu}.$$

Observe that by (9.15)

$$\limsup_{t\to\infty}\frac{B(0,t)}{\lambda_1^t}\le\frac{x_{10}+x_{11}}{x_{10}}B\quad\text{a.s.},$$

$$\limsup_{t\to\infty}\frac{B(1,t)}{\lambda_1^t}\le\frac{x_{10}+x_{11}}{x_{11}}B\quad\text{a.s.}$$

and by (9.14)

$$\lim_{\nu\to\infty}\frac{m_{01}^{(\nu)}-x_{10}y_{11}\lambda_1^\nu}{\lambda_1^\nu}=\lim_{\nu\to\infty}\frac{(\lambda_1^\nu-\lambda_0^\nu)x_{10}y_{11}-x_{10}y_{11}\lambda_1^\nu}{\lambda_1^\nu}=0,$$

$$\lim_{\nu\to\infty}\frac{m_{11}^{(\nu)}-x_{11}y_{11}\lambda_1^\nu}{\lambda_1^\nu}=\lim_{\nu\to\infty}\frac{(\lambda_1^\nu-\lambda_0^\nu)x_{11}y_{11}+\lambda_0^\nu-x_{11}y_{11}\lambda_1^\nu}{\lambda_1^\nu}=0.$$

Consequently for any $\varepsilon>0$ there exists a $\nu=\nu(\varepsilon)$ such that

$$\limsup_{t\to\infty}\left|\frac{m_{01}^{(\nu)}B(0,t)+m_{11}^{(\nu)}B(1,t)}{\lambda_1^{t+\nu}}-\frac{x_{10}B(0,t)+x_{11}B(1,t)}{\lambda_1^t}y_{11}\right|\le\varepsilon\quad\text{a.s.}$$

Hence by (9.19) and (9.20) we have (iv) and, in turn, Theorem 9.1.

Note that in case $I(M)=0$, $\det M=0$ we have

$$\begin{pmatrix}m_{00} & m_{01}\\ \mu m_{00} & \mu m_{01}\end{pmatrix}\begin{pmatrix}1\\ \mu\end{pmatrix}=(m_{00}+\mu m_{01})\begin{pmatrix}1\\ \mu\end{pmatrix}.$$

Hence denoting $m_{00}+\mu m_{01}$ by λ_1, Theorem 9.1 can be reformulated as follows.

THEOREM 9.2 *Assume that* $I(M)=0$.

(i) *If* $\lambda_1\le 1$ *then*

$$\lim_{t\to\infty}B(0,t)=\lim_{t\to\infty}B(1,t)=0\qquad a.s.$$

(ii) *If* $\lambda_1>1$ *then there exists a r.v.* $B\ge 0$ *with* $\mathbf{E}B=1$ *such that*

$$\lim_{t\to\infty}\frac{B(0,t)}{\lambda_1^t}=C_0(M)B\qquad a.s.$$

and

$$\lim_{t\to\infty}\frac{B(1,t)}{\lambda_1^t}=C_1(M)B\qquad a.s.$$

where

$$C_0(M) = \begin{cases} m_{00}(1+\mu) & \text{if} \quad \det M = 0, \\ -\dfrac{(x_{10}+x_{11})x_{01}}{\det \Xi} & \text{if} \quad \det M \neq 0, \end{cases}$$

$$C_1(M) = \begin{cases} m_{01}(1+\mu) & \text{if} \quad \det M = 0, \\ \dfrac{(x_{10}+x_{11})x_{00}}{\det \Xi} & \text{if} \quad \det M \neq 0. \end{cases}$$

The following two theorems are trivial. Their proofs will be omitted.

THEOREM 9.3

$$\lim_{t\to\infty} B(0,t) = \lim_{t\to\infty} B(1,t) = 0 \qquad a.s.$$

in the following cases:

(i) $I(M) = 4$,

(ii) $I(M) = 3$ *and* $m_{00} + m_{11} \leq 1$,

(iii) $I(M) = 2$, $m_{00} \leq 1$, $m_{11} \leq 1$ *and* $m_{01}m_{10} \leq 1$,

(iv) $I(M) = 1$, $m_{01}m_{10} = 0$, $m_{00} \leq 1$ *and* $m_{11} \leq 1$.

THEOREM 9.4 (i) *If* $I(M) = 3$ *and* $m_{00} > 1$ *resp.* $m_{11} > 1$ *then there exist r.v.'s* $B_0 \geq 0$ *resp.* $B_1 \geq 0$ *with* $\mathbf{E}B_0 = \mathbf{E}B_1 = 1$ *such that*

$$\lim_{t\to\infty} \frac{B(0,t)}{m_{00}^t} = B_0 \quad and \quad \lim_{t\to\infty} B(1,t) = 0 \quad a.s.$$

resp.

$$\lim_{t\to\infty} \frac{B(1,t)}{m_{11}^t} = B_1 \quad and \quad \lim_{t\to\infty} B(0,t) = 0 \quad a.s.$$

(ii) *If* $I(M) = 2$, $m_{00} > 1$ *and* $m_{01} > 0$ *then there exists a r.v.* $B \geq 0$ *with* $\mathbf{E}B = 1$ *such that*

$$\lim_{t\to\infty} \frac{B(0,t)}{m_{00}^t} = \lim_{t\to\infty} \frac{B(1,t)}{m_{01}m_{00}^{t-1}} = B$$

(iii) *If* $I(M) = 2$, $m_{01}+m_{10} = 0$, $m_{00} > 1$ *and* $m_{11} > 1$ *then there exist independent r.v.'s* $B_0 \geq 0$ *and* $B_1 \geq 0$ *with* $\mathbf{E}B_0 = \mathbf{E}B_1 = 1$ *such that*

$$\lim_{t\to\infty} \frac{B(0,t)}{m_{00}^t} = B_0 \quad and \quad \lim_{t\to\infty} \frac{B(1,t)}{m_{11}^t} = B_1 \qquad a.s.$$

(iv) *If* $I(M) = 2$, $m_{01} + m_{10} = 0$, $m_{00} > 1$ *and* $m_{11} \leq 1$ *then there exists a r.v.* $B \geq 0$ *with* $\mathbf{E}B = 1$ *such that*

$$\lim_{t\to\infty} \frac{B(0,t)}{m_{00}^t} = B \quad and \quad \lim_{t\to\infty} B(1,t) = 0 \qquad a.s.$$

(v) *If* $I(M) = 2$, $m_{01} + m_{10} = 0$, $m_{00} \leq 1$ *and* $m_{11} > 1$ *then there exists a r.v.* $B \geq 0$ *with* $\mathbf{E}B = 1$ *such that*

$$\lim_{t\to\infty} \frac{B(1,t)}{m_{11}^t} = B \quad \text{and} \quad \lim_{t\to\infty} B(0,t) = 0 \qquad a.s.$$

(vi) *If* $I(M) = 2$, $m_{00} + m_{11} = 0$ *and* $m_{01}m_{10} > 1$ *then there exist two independent r.v.'s* $B_0 \geq 0$ *and* $B_1 \geq 0$ *with* $\mathbf{E}B_0 = \mathbf{E}B_1 = 1$ *such that*

$$\lim_{t\to\infty} \frac{B(0,2t)}{(m_{01}m_{10})^t} = \lim_{t\to\infty} \frac{B(1,2t+1)}{(m_{01}m_{10})^t m_{01}} = B_0 \qquad a.s.$$

and

$$\lim_{t\to\infty} \frac{B(1,2t)}{(m_{01}m_{10})^t} = \lim_{t\to\infty} \frac{B(0,2t+1)}{(m_{01}m_{10})^t m_{10}} = B_1 \qquad a.s.$$

(vii) *If* $I(M) = 2$, $m_{01} + m_{11} = 0$ *and* $m_{00} > 1$ *then there exists a r.v.* $B \geq 0$ *with* $\mathbf{E}B = 1$ *such that*

$$\lim_{t\to\infty} \frac{B(0,t)}{m_{00}^t} = B \quad \text{and} \quad \lim_{t\to\infty} B(1,t) = 0 \qquad a.s.$$

(viii) *If* $I(M) = 2$, $m_{11} > 1$ *and* $m_{10} > 0$ *then there exists a r.v.* $B \geq 0$ *with* $\mathbf{E}B = 1$ *such that*

$$\lim_{t\to\infty} \frac{B(0,t)}{m_{11}^{t-1}m_{10}} = \lim_{t\to\infty} \frac{B(1,t)}{m_{11}^t} = B \qquad a.s.$$

(ix) *If* $I(M) = 1$, $m_{01} = 0$ *and* $m_{00} > m_{11} > 1$ *resp.* $m_{10} = 0$ *and* $m_{11} > m_{00} > 1$ *then there exist two independent r.v.'s* $B_0 \geq 0$, $B_1 \geq 0$ *with* $\mathbf{E}B_0 = \mathbf{E}B_1 = 1$ *such that*

$$\lim_{t\to\infty} \frac{B(1,t)}{m_{11}^t} = B_1 \qquad a.s.$$

and

$$\lim_{t\to\infty} \frac{B(0,t)}{m_{00}^t} = B_0 \left(1 + \frac{m_{10}}{m_{00} - m_{11}}\right) \qquad a.s.$$

resp.

$$\lim_{t\to\infty} \frac{B(0,t)}{m_{00}^t} = B_0 \qquad a.s.$$

and

$$\lim_{t\to\infty} \frac{B(1,t)}{m_{11}^t} = B_1 \left(1 + \frac{m_{01}}{m_{11} - m_{00}}\right) \qquad a.s.$$

(x) *If* $I(M) = 1$, $m_{01} = 0$ *and* $m_{00} > 1 \geq m_{11}$ *resp.* $m_{10} = 0$ *and* $m_{11} > 1 \geq m_{00}$ *then there exists a r.v.* $B \geq 0$ *with* $\mathbf{E}B = 1$ *such that*

$$\lim_{t\to\infty} \frac{B(0,t)}{m_{00}^t} = B \quad \text{and} \quad \lim_{t\to\infty} B(1,t) = 0 \qquad a.s.$$

resp.

$$\lim_{t\to\infty} \frac{B(1,t)}{m_{11}^t} = B \quad \text{and} \quad \lim_{t\to\infty} B(0,t) = 0 \qquad a.s.$$

Going through on the above results we can observe that they cover all possible case except the cases

(i) $I(M) = 1,\ m_{01} = 0,\ m_{11} \geq m_{00}$ and $\max(m_{11}, m_{00}) > 1$,

(ii) $I(M) = 1,\ m_{10} = 0,\ m_{00} \geq m_{11}$ and $\max(m_{11}, m_{00}) > 1$.

These cases are more delicate and will be settled in the following:

THEOREM 9.5 *Let* $I(M) = 1$.

(i) *If* $m_{01} = 0,\ m_{11} > m_{00}$ *and* $m_{11} > 1$ *resp.* $m_{10} = 0,\ m_{00} > m_{11}$ *and* $m_{00} > 1$ *then there exist two r.v.'s* $B \geq 0$ *resp.* $B' \geq 0$ *with* $\mathbf{E}B = \mathbf{E}B' = 1$ *such that*

$$\lim_{t\to\infty} \frac{B(1,t)}{m_{11}^t} = \lim_{t\to\infty}(m_{11} - m_{00})\frac{B(0,t)}{m_{11}^t m_{10}} = B \qquad a.s.$$

resp.

$$\lim_{t\to\infty} \frac{B(0,t)}{m_{00}^t} = \lim_{t\to\infty}(m_{00} - m_{11})\frac{B(1,t)}{m_{00}^t m_{01}} = B' \qquad a.s.$$

(ii) *If* $m_{01} = 0$ *resp.* $m_{10} = 0$ *and* $m_{00} = m_{11} > 1$ *then there exist two r.v.'s* $B \geq 0$ *resp.* $B' \geq 0$ *with* $\mathbf{E}B = \mathbf{E}B' = 1$ *such that*

$$\lim_{t\to\infty} \frac{B(1,t)}{m_{11}^t} = \lim_{t\to\infty} \frac{B(0,t)}{m_{10}m_{11}^{t-1}t} = B \qquad a.s.$$

resp.

$$\lim_{t\to\infty} \frac{B(0,t)}{m_{11}^t} = \lim_{t\to\infty} \frac{B(0,t)}{m_{11}^t} = \lim_{t\to\infty} \frac{B(1,t)}{m_{01}m_{11}^{t-1}t} = B' \qquad a.s.$$

Proof. Let $m_{01} = 0$. Then clearly we have

$$\lim_{t\to\infty} \frac{B(1,t)}{m_{11}^t} = B \qquad \text{a.s.} \tag{9.21}$$

By the law of large numbers for any integer $\nu \geq 1$ we have

$$\lim_{t\to\infty} \frac{B(0,t+\nu)}{A(t,\nu)} = 1 \qquad \text{a.s.}$$

where

$$A(t,\nu) = m_{00}^{\nu}B(0,t) + \sum_{j=1}^{\nu} m_{00}^{\nu-j}m_{10}B(1,t+j-1).$$

By (9.21) as $t \to \infty$ we have

$$A(t,\nu) \sim m_{00}^{\nu}B(0,t) + Bm_{10}m_{11}^{t+\nu-1}\sum_{j=1}^{\nu}\left(\frac{m_{00}}{m_{11}}\right)^{\nu-j}.$$

Observe that if ν is large enough then

$$\sum_{j=1}^{\nu}\left(\frac{m_{00}}{m_{11}}\right)^{\nu-j} \sim \begin{cases} \nu & \text{if} \quad m_{00}=m_{11}, \\ \dfrac{m_{11}}{m_{11}-m_{00}} & \text{if} \quad m_{11}>m_{00} \end{cases}$$

which easily implies the Theorem in case $m_{01}=0$. The case $m_{10}=0$ can be treated similarly.

9.2 The model

At time $t=0$ two particles of different types – type 0 and type 1 – are located in $0 \in \mathbb{Z}^d$ and they begin independent random walks. At time $t=1$ they produce particles by the procedure described in Section 9.1 and the offsprings execute independent random walks just like in Section 4.2.

Let $\lambda_i(x,t)$ $(i=0,1,\ x \in \mathbb{Z}^d,\ t=0,1,2,\dots)$ be the number of particles of type i located in x at time t. We are interested in the limit properties of $\{\lambda_i(x,t)\ i=0,1,\ x\in \mathbb{Z}^d\}$ as $t \to \infty$.

At first we give a formal definition of $\lambda_i(x,t)$. Let

$$\{X(i,x,t,\mu),(Z(i,0,x,t,\mu),Z(i,1,x,t,\mu))\}$$

$(x \in \mathbb{Z}^d,\ t=0,1,2,\dots,\ \mu=1,2,\dots,\ i=0,1)$ be an array of independent r.v.'s with

$$\mathbf{P}\{X(i,x,t,\mu)=e_k\}=\mathbf{P}\{X(i,x,t,\mu)=-e_k\}=\frac{1}{2d}$$

$(k=1,2,\dots,d)$,

$$\mathbf{P}\{Z(i,0,x,t,\mu)=k,\ Z(i,1,x,t,\mu)=\ell\}=p(i,k,\ell)$$

where $p(i,k,\ell)$ satisfies the conditions of Section 9.1.

Let

$$\lambda_0(x,0)=\lambda_1(x,0)=\begin{cases}1 & \text{if} \quad x=0, \\ 0 & \text{if} \quad x\neq 0\end{cases}$$

and

$$\lambda_j(x,t)=\sum_{y\in C(x)}\sum_{i=0}^{1}\sum_{\mu=1}^{\lambda_i(y,t-1)} I_{x-y}(X(i,y,t-1,\mu))Z(i,j,y,t-1,\mu)$$

where $x\in\mathbb{Z}^d$, $t=1,2,\dots,$ $i=0,1$ and $j=0,1$.

The intuitive meaning of the above definition is the following. In order to get the number of the type j $(j=0,1)$ particles located in x at time t, consider the type 0 and the type 1 particles located in one of the neighbours of x at time $t-1$. The μ-th type i particle located in $y\in C(x)$ at time $t-1$ is moving to x if

$X(i,y,t-1,\mu) = x - y$ i.e. if $I_{x-y}(X(i,y,t-1,\mu)) = 1$. If this particle moves to x then it will produce there

$$Z(i,j,y,t-1,\mu) = I_{x-y}(X(i,y,t-1,\mu))Z(i,j,y,t-1,\mu)$$

type j offsprings.

The limit properties of $\lambda_i(x,t)$ when

$$\lim_{t\to\infty} B(0,t) = 0 \quad \text{and/or} \quad \lim_{t\to\infty} B(1,t) = 0$$

are trivial. Hence we do not study the cases of (i), (iii) of Theorem 9.1 (resp. (i) of Theorem 9.2). Similarly the study of all cases of Theorem 9.3 will be omitted.

By Theorem 4.2 the limit behaviour of $\lambda_i(x,t)$ can be described easily in the following cases: (i), (iv), (v), (vii) and (x) of Theorem 9.4.

The cases of (iii) and (vi) of Theorem 9.4 can be also easily settled. In fact we have

THEOREM 9.6 (i) *If $I(M) = 2$, $m_{01} + m_{10} = 0$, $m_{00} > 1$ and $m_{11} > 1$ then there exist two independent r.v.'s $B_0 \geq 0$ and $B_1 \geq 0$ with $\mathbf{E}B_0 = \mathbf{E}B_1 = 1$ such that*

$$\lim_{T\to\infty} T^{1-\varepsilon}\left|\frac{1}{2}\left(\frac{2\pi T}{d}\right)^{d/2}\frac{\lambda_i(x,T)}{m_{ii}^T} - B_i\right| = 0 \qquad a.s. \quad (i = 0,1)$$

for any fixed $x \in \mathbb{Z}^d$ and $0 < \varepsilon < 1$ provided that $x \equiv T \pmod 2$.

(ii) *If $I(M) = 2$, $m_{00} + m_{11} = 0$ and $m_{01}m_{10} > 1$ then there exist two independent r.v.'s $B_0 \geq 0$ and $B_1 \geq 0$ with $\mathbf{E}B_0 = \mathbf{E}B_1 = 1$ such that*

$$\begin{aligned}
&\lim_{T\to\infty}\left|\frac{1}{2}\left(\frac{4\pi T}{d}\right)^{d/2}\frac{\lambda_i(x,2T)}{(m_{01}m_{10})^T} - B_i\right| = \\
&= \lim_{T\to\infty} T^{1-\varepsilon}\left|\frac{1}{2}\left(\frac{4\pi T + 2\pi}{d}\right)^{d/2}\frac{\lambda_0(y,2T+1)}{(m_{01}m_{10})^T m_{10}} - B_1\right| = \\
&= \lim_{T\to\infty} T^{1-\varepsilon}\left|\frac{1}{2}\left(\frac{4\pi T + 2\pi}{d}\right)^{d/2}\frac{\lambda_1(y,2T+1)}{(m_{01}m_{10})^T m_{01}} - B_0\right| = 0 \qquad a.s.
\end{aligned}$$

for any fixed $x \in \mathbb{Z}^d$ and $0 < \varepsilon < 1$ provided that $x \equiv 0 \pmod 2$ and $y \equiv 1 \pmod 2$.

Proof is trivial by Theorem 9.4 (cases (iii) and (vi)) and Theorem 4.8 (4.31).

Hence we are interested only in the following cases:

(ii) of Theorems 9.2,

(ii), (viii), (ix) of Theorem 9.4,

(i), (ii) of Theorem 9.5.

9.3 On the moments of $\lambda_i(x,t)$

Introduce the following notations:

(i) let $\{S_t;\ t = 0,1,2,\dots\}$ be a simple, symmetric nearest neighbour random walk on $\mathbb{Z}^d$ with $S_0 = 0$,

(ii)
$$p(u \rightsquigarrow v, t) = \mathbf{P}\{S_{s+t} = v \mid S_s = u\},$$

(iii)
$$f_i(x,t,T) = \\ = m_{0i}^{(T-t)} \sum_{y\in\mathbb{Z}^d} \lambda_0(y,t)p(y \rightsquigarrow x, T-t) + m_{1i}^{(T-t)} \sum_{y\in\mathbb{Z}^d} \lambda_1(y,t)p(y \rightsquigarrow x, T-t)$$

where $i = 0,1;\ 0 \le t \le T$,

(iv) let
$$\mathcal{F}(T) = \mathcal{F}\{\lambda_i(x,t);\ x \in \mathbb{Z}^d,\ t = 0,1,2,\dots,T,\ i = 0,1\}$$

be the smallest σ–algebra with respect to which the array
$$\{\lambda_i(x,t);\ x \in \mathbb{Z}^d,\ t = 0,1,2,\dots,T,\ i = 0,1\}$$

is measurable,

(v)
$$\sum_{y\in\mathbb{Z}^d} \lambda_i(y,t) = B(i,t) \qquad (i = 0,1).$$

At first we present a number of lemmas.

LEMMA 9.3 *For any $0 \le t \le T$, $x \in \mathbb{Z}^d$ and $i = 0,1$ we have*
$$\mathbf{E}(\lambda_i(x,T) \mid \mathcal{F}(t)) = f_i(x,T,t). \tag{9.22}$$

Proof. Since
$$\mathbf{E}(\lambda_j(x,t) \mid \mathcal{F}(t)) = \lambda_j(x,t) \qquad (j = 0,1),$$
$m_{00}^{(0)} = m_{11}^{(0)} = 1,\ m_{01}^{(0)} = m_{10}^{(0)} = 0$, we have
$$f_j(x,T,T) = \sum_{y\in\mathbb{Z}^d} \lambda_j(y,T)p(y \rightsquigarrow x, 0) = \lambda_j(x,T).$$

Consequently we have (9.22) in case $T = t$. It is also easy to see that (9.22) holds if $T = 1$ and $t = 0$. Now we give the proof by induction. Assume that (9.22) holds true for T, for any $0 \le t \le T$ and $x \in \mathbb{Z}^d$. Then

$$\begin{aligned}
&\mathbf{E}(\lambda_j(x, T+1) \mid \mathcal{F}(t)) = \\
&= \mathbf{E}\left(\sum_{y \in C(x)} \sum_{i=0}^{1} \sum_{\mu=1}^{\lambda_i(y,T)} I_{x-y}(X(i,y,T,\mu)) Z(i,j,y,T,\mu) \mid \mathcal{F}(t) \right) = \\
&= \sum_{i=0}^{1} \frac{m_{ij}}{2d} \sum_{y \in C(x)} \mathbf{E}(\lambda_i(y,T) \mid \mathcal{F}(t)) = \sum_{i=0}^{1} \frac{m_{ij}}{2d} \sum_{y \in C(x)} f_i(y,T,t) = \\
&= \sum_{i=0}^{1} \frac{m_{ij}}{2d} \sum_{y \in C(x)} \sum_{\ell=0}^{1} m_{\ell i}^{T-t} \sum_{u \in \mathbb{Z}^d} \lambda_\ell(u,t) p(u \rightsquigarrow y, T-t) = \\
&= \frac{1}{2d} \sum_{\ell=0}^{1} m_{\ell j}^{(T+1-t)} \sum_{u \in \mathbb{Z}^d} \lambda_\ell(u,t) \sum_{y \in C(x)} p(u \rightsquigarrow y, T-t) = \\
&= \sum_{\ell=0}^{1} m_{\ell j}^{(T+1-t)} \sum_{u \in \mathbb{Z}^d} \lambda_\ell(u,t) p(u \rightsquigarrow x, T-t) = f_j(x,T,t).
\end{aligned}$$

Hence Lemma 9.3 is proved.

LEMMA 9.4

$$\mathbf{E}(f_i(x,T,t) \mid \mathcal{F}(t-1)) = f_i(x,T,t-1), \tag{9.23}$$

$$\mathbf{E}\lambda_i(x,T) = (m_{0i}^{(T)} + m_{1i}^{(T)}) p(0 \rightsquigarrow x, T) = \mathbf{E} f_i(x,T,t) = p(0 \rightsquigarrow x, T) \mathbf{E} B_i(T). \tag{9.24}$$

Proof. By (9.22) we have

$$\begin{aligned}
\mathbf{E}(f_i(x,T,t) \mid \mathcal{F}(t-1)) &= \mathbf{E}(\mathbf{E}(\lambda_i(x,T) \mid \mathcal{F}(t)) \mid \mathcal{F}(t-1)) = \\
&= \mathbf{E}(\lambda_i(x,T) \mid \mathcal{F}(t-1)) = f_i(x,T,t-1)
\end{aligned}$$

and (9.23) is proved. By (9.1), (9.2) and (9.22) we have (9.24).

LEMMA 9.5 *Let*

$$\mathbf{E} Z(i,0,x,t,\mu) Z(i,1,x,t,\mu) = \sum_{k=0}^{\infty} \sum_{\ell=0}^{\infty} k \ell p(i,k,\ell) = \rho(i) \quad (i = 0,1).$$

Then we have

$$\begin{aligned}
&\sum_{x \in \mathbb{Z}^d} \mathbf{E}((f_i(x,T,t) - f_i(x,T,t-1))^2 \mid \mathcal{F}(t-1)) \sim \\
&\sim 2 \left(\frac{d}{4\pi} \right)^{d/2} \frac{1}{(T-t)^{d/2}} \sum_{j=0}^{1} \sum_{\ell=0}^{1} (m_{ji}^{(T-t)})^2 (\sigma_{\ell j}^2 + \rho(\ell) - m_{\ell 0} m_{\ell 1}) B_\ell(t-1).
\end{aligned}$$

Proof. Observe that

$$f_i(x,T,t) = \sum_{j=0}^{1} m_{ji}^{(T-t)} \sum_{y\in\mathbb{Z}^d} \lambda_j(y,t) p(y \rightsquigarrow x, T-t)$$

where

$$\lambda_j(y,t) = \sum_{u\in C(y)} \sum_{\ell=0}^{1} \sum_{\mu=1}^{\lambda_\ell(u,t-1)} V(\ell,j,u,y,t,\mu)$$

and

$$V = V(\ell,j,u,y,t,\mu) = I_{y-u}(X(\ell,u,t-1,\mu)) Z(\ell,j,u,t-1,\mu).$$

Hence

$$\begin{aligned}
&f_i(x,T,t) = \\
&= \sum_{j=0}^{1} m_{ji}^{(T-t)} \sum_{y\in\mathbb{Z}^d} \sum_{u\in C(y)} \sum_{\ell=0}^{1} \sum_{\mu=1}^{\lambda_\ell(u,t-1)} V p(y \rightsquigarrow x, T-t) = \\
&= \sum_{j=0}^{1} \sum_{\ell=0}^{1} m_{ji}^{(T-t)} \sum_{y\in C(u)} \sum_{u\in\mathbb{Z}^d} \sum_{\mu=1}^{\lambda_\ell(u,t-1)} V p(y \rightsquigarrow x, T-t) = \\
&= \sum_{j=0}^{1} \sum_{\ell=0}^{1} m_{ji}^{(T-t)} \sum_{u\in C(y)} \sum_{y\in\mathbb{Z}^d} \sum_{\mu=1}^{\lambda_\ell(y,t-1)} V(\ell,j,y,u,t,\mu) p(u \rightsquigarrow x, T-t).
\end{aligned}$$

Similarly

$$\begin{aligned}
&f_i(x,T,t-1) = \\
&= \sum_{\ell=0}^{1} m_{\ell i}^{(T-t+1)} \sum_{y\in\mathbb{Z}^d} \lambda_\ell(y,t-1) p(y \rightsquigarrow x, T-t+1) = \\
&= \sum_{\ell=0}^{1} \sum_{j=0}^{1} m_{\ell j} m_{ji}^{(T-t)} \sum_{y\in\mathbb{Z}^d} \sum_{\mu=1}^{\lambda_\ell(y,t-1)} p(y \rightsquigarrow x, T-t+1).
\end{aligned}$$

Consequently

$$f_i(x,T,t) - f_i(x,T,t-1) = \sum_{j=0}^{1} \sum_{\ell=0}^{1} m_{ji}^{(T-t)} \sum_{y\in\mathbb{Z}^d} \sum_{\mu=1}^{\lambda_\ell(y,t-1)} Q$$

where

$$\begin{aligned}
&Q = Q(\ell,j,y,x,\mu,T,t) = \\
&= \sum_{u\in C(y)} V(\ell,j,y,u,t,\mu) p(u \rightsquigarrow x, T-t) - m_{\ell j} p(y \rightsquigarrow x, T-t+1) = \\
&= Z(\ell,j,y,t-1,\mu) \sum_{u\in C(y)} I_{u-y}(X(\ell,y,t-1,\mu)) p(u \rightsquigarrow x, T-t) - \\
&\quad - m_{\ell j} p(y \rightsquigarrow x, T-t+1).
\end{aligned}$$

Following the proof of Lemma 4.6 we have

$$\mathbf{E}(Q^2 \mid \mathcal{F}(t-1)) = \mathbf{E}Q^2 =$$
$$= (m_{\ell j}^2 + \sigma_{\ell j}^2)\frac{1}{2d}\sum_{u\in C(y)} p^2(u \rightsquigarrow x, T-t) - m_{\ell j}^2 p^2(y \rightsquigarrow x, T-t+1)$$

and we get similarly

$$\mathbf{E}Q(\ell,0,\ldots)Q(\ell,1,\ldots) = \mathbf{E}(Q(\ell,0,\ldots)Q(\ell,1,\ldots) \mid \mathcal{F}(t-1)) =$$
$$= \rho(\ell)\frac{1}{2d}\sum_{u\in C(y)} p^2(u \rightsquigarrow x, T-t) - m_{\ell 0}m_{\ell 1}p^2(y \rightsquigarrow x, T-t+1).$$

Hence we have

$$\mathbf{E}((f_i(x,T,t) - f_i(x,T,t-1))^2 \mid \mathcal{F}(t-1)) =$$
$$= \sum_{j=0}^{1}\sum_{\ell=0}^{1}(m_{ji}^{(T-y)})^2 \sum_{y\in\mathbb{Z}^d} \lambda_\ell(y,t-1)\Delta + \sum_{\ell=0}^{1}\sum_{j=0}^{1}(m_{ji}^{(T-t)})^2 \sum_{y\in\mathbb{Z}^d}\lambda_\ell(y,t-1)\Delta^* =$$
$$= \sum_{j=0}^{1}\sum_{\ell=0}^{1}(m_{ji}^{(T-t)})^2 \sum_{y\in\mathbb{Z}^d}\lambda_\ell(y,t-1)(\Delta+\Delta^*)$$

where

$$\Delta = \frac{m_{\ell j}^2 + \sigma_{\ell j}^2}{2d}\sum_{u\in C(y)} p^2(u \rightsquigarrow x, T-t) - m_{\ell j}^2 p^2(y \rightsquigarrow x, T-t+1)$$

and

$$\Delta^* = \frac{\rho(\ell)}{2d}\sum_{u\in C(y)} p^2(u \rightsquigarrow x, T-t) - m_{\ell 0}m_{\ell 1}p^2(y \rightsquigarrow x, T-t+1).$$

Note that by Lemma 4.2

$$\sum_{x\in\mathbb{Z}^d}\Delta = (m_{\ell j}^2 + \sigma_{\ell j}^2)p(0 \rightsquigarrow 0, 2T-2t) - m_{\ell j}^2 p(0 \rightsquigarrow 0, 2T-2t+2) \sim$$
$$\sim 2\sigma_{\ell j}^2\left(\frac{d}{4\pi}\right)^{d/2}\frac{1}{(T-t)^{d/2}}$$

and

$$\sum_{x\in\mathbb{Z}^d}\Delta^* = \rho(\ell)p(0 \rightsquigarrow 0, 2T-2t) - m_{\ell 0}m_{\ell 1}p(0 \rightsquigarrow 0, 2T-2t+2) \sim$$
$$\sim 2(\rho(\ell) - m_{\ell 0}m_{\ell 1})\left(\frac{d}{4\pi}\right)^{d/2}\frac{1}{(T-t)^{d/2}}.$$

Consequently

$$\sum_{x\in\mathbb{Z}^d} \mathbf{E}((f_i(x,T,t) - f_i(x,T,t-1))^2 \mid \mathcal{F}(t-1)) =$$

$$= \sum_{j=0}^{1}\sum_{\ell=0}^{1}(m_{ji}^{(T-t)})^2 \sum_{y\in\mathbb{Z}^d} \lambda_\ell(y,t-1) \sum_{x\in\mathbb{Z}^d} (\Delta + \Delta^*)$$

which implies Lemma 9.5.

Assuming the conditions of Lemma 9.2 by (9.14) we have

$$\sum_{j=0}^{1}(m_{ji}^{(t)})^2 \sim \lambda_1^{2t} y_{i1}^2 (x_{10}^2 + x_{11}^2).$$

Without loss of generality we might assume

$$x_{10}^2 + x_{11}^2 = x_{00}^2 + x_{01}^2 = 1.$$

Hence Lemma 9.5 can be reformulated as follows.

LEMMA 9.6 *Assume the conditions of* (iv) *of Theorem* 9.1. *Then*

$$\sum_{x\in\mathbb{Z}^d} \mathbf{E}((f_i(x,T,t) - f_i(x,T,t-1))^2 \mid \mathcal{F}(t-1)) \sim$$

$$\sim 2\left(\frac{d}{4\pi}\right)^{d/2} \frac{\lambda_1^{2T-2t}}{(T-t)^{d/2}} y_{i1}^2 \sum_{j=0}^{1}\sum_{\ell=0}^{1}(\sigma_{\ell j}^2 + \rho(\ell) - m_{\ell 0}m_{\ell 1})B(\ell,t-1).$$

Repeating the proofs of Lemmas 4.7 and 4.8 we have

LEMMA 9.7 *Assume that $T-t\to\infty$ and the conditions of Lemma 9.6 are fulfilled. Then we have*

$$\sum_{x\in\mathbb{Z}^d} \mathbf{E}((\lambda_i(x,T) - f_i(x,T,t))^2 \mid \mathcal{F}(t)) \sim$$

$$\sim 2\left(\frac{d}{4\pi}\right)^{d/2} \sum_{j=0}^{1}\sum_{\ell=0}^{1}(\sigma_{\ell j}^2 + \rho(\ell) - m_{\ell 0}m_{\ell 1})B(\ell,t-1) \sum_{k=0}^{T-t} k^{-d/2}\lambda_1^{T+i-t-2} \le$$

$$\le C\frac{1}{\lambda_1^t (T-t)^{d/2}} \frac{B(0,t)+B(1,t)}{\lambda_1^t}$$

with an absolute constant $C>0$.

THEOREM 9.7 *Assume that $I(M)=0$, $\det M \neq 0$ and $\lambda_1 > 1$. Then for any $0\le\alpha\le 1$ and $i=0,1$ we have*

$$\mathbf{E}\left(\sum_{x\in C(0,T^\alpha)}\left|\frac{\lambda_i(x,T)}{\lambda_1^T} - \frac{f_i(x,T,t)}{\lambda_1^T}\right| \,\middle|\, \mathcal{F}(t)\right) \le$$

$$\le C(2T^\alpha)^{d/2}\left(\frac{1}{\lambda_1^t(T-t)^{d/2}}\frac{B(0,t)+B(1,t)}{\lambda_1^t}\right)^{1/2}$$

and

$$\mathbf{E}\sum_{x\in C(0,T^{\alpha})}\left|\frac{\lambda_i(x,T)}{\lambda_1^T}-\frac{f_i(x,T,t)}{\lambda_1^T}\right|\le C(2T^{\alpha})^{d/2}(\lambda_1^t(T-t)^{d/2})^{-1/2}.$$

Proof is the same as that of Theorem 4.3.

9.4 Global limit theorems

In this and the next Sections we give the analogues of the results of Sections 4.4 and 4.5 in case $I(M)=0$, $\det M\neq 0$, $\lambda_1>1$ without proofs.

LEMMA 9.8 (cf. Lemma 4.9)

$$\begin{aligned}&(m_{0i}^{(T-t)}B(0,t)+m_{1i}^{(T-t)}B(1,t))\inf_{y\in D(t)}p(0\rightsquigarrow x-y,T-t)\le\\&\le\mathbf{E}(\lambda_i(x,T)\mid\mathcal{F}(t))\le(m_{0i}^{(T-t)}B(0,t)+m_{1i}^{(T-t)}B(1,t))\sup_{y\in D(t)}p(0\rightsquigarrow x-y,T-t).\end{aligned}$$

Note that by (9.14) as $T-t\to\infty$

$$\frac{m_{0i}^{(T-t)}}{\lambda_1^{T-t}}\to x_{10}y_{i1}\quad\text{and}\quad\frac{m_{1i}^{(T-t)}}{\lambda_1^{T-t}}\to x_{11}y_{i1}.$$

Hence by (9.15) and (iv) of Theorem 9.1 as $T-t\to\infty$ and $t\to\infty$ we have

$$\begin{aligned}&\frac{m_{0i}^{(T-t)}B(0,t)+m_{i1}^{(T-t)}B(1,t)}{\lambda_1^T}\sim\\&\sim y_{i1}\frac{x_{10}B(0,t)+x_{11}B(1,t)}{\lambda_1^t}=y_{i1}M_1(t)\to y_{i1}(x_{10}+x_{11})B=\\&=\begin{cases}\dfrac{-(x_{10}+x_{11})x_{01}}{\det\Xi}B & \text{if}\quad i=0,\\[2mm] \dfrac{(x_{10}+x_{11})x_{00}}{\det\Xi}B & \text{if}\quad i=1\end{cases}=\lim_{t\to\infty}\frac{B(i,t)}{\lambda_1^t}.\end{aligned}$$

THEOREM 9.8 (cf. Theorems 4.4 and 4.5) *For any $0<\varepsilon<1/2$ there exists a $C=C(\varepsilon)>0$ such that for any $T=1,2,\ldots$ and $i=0,1$ we have*

$$\mathbf{E}\sum_{x\in\mathbb{Z}^d}\left|\frac{\lambda_i(x,T)}{\lambda_1^T}-p(0\rightsquigarrow x,T)B_i\right|\le CT^{-(1/2-\varepsilon)}$$

and

$$\lim_{T\to\infty}T^{1/2-\varepsilon}\sum_{x\in\mathbb{Z}^d}\left|\frac{\lambda_i(x,T)}{\lambda_1^T}-p(0\rightsquigarrow x,T)B_i\right|=0\quad a.s.$$

where

$$B_i=\lim_{t\to\infty}\frac{B(i,t)}{\lambda_1^t}.$$

9.5 Local limit theorems

THEOREM 9.9 (cf. Theorem 4.7) *For any $0 \le \gamma \le 1/2$, $\varepsilon > 0$ and $i = 0,1$ we have*

$$\mathbf{E} \sum_{x \in D(t^\gamma)} \left| \frac{\lambda_i(x,T)}{\lambda_1^T} - p(0 \leadsto x, T) B_i \right| \le C T^{-(d+2-2\gamma(d+1)-\varepsilon)/2}$$

if C is big enough and

$$\lim_{T\to\infty} T^{(d+2-2\gamma(d+1)-\varepsilon)/2} \sum_{x \in D(T^\gamma)} \left| \frac{\lambda_i(x,T)}{\lambda_1^T} - p(0 \leadsto x, T) B_i \right| = 0 \quad a.s.$$

THEOREM 9.10 (cf. Theorem 4.8) *Let $x = x(T) \in D(T^\gamma)$ $(0 \le \gamma \le 1)$ be a sequence of vectors. Then for any $0 < \varepsilon < \gamma$ and $i = 0,1$ we have*

$$\mathbf{P}\left\{ T^{(d+2-2\gamma-2\varepsilon)/2} \left| \frac{\lambda_i(x,T)}{\lambda_1^T} - p(0 \leadsto x, T) B_i \right| \ge 1 \right\} \le \exp(-O(T^\varepsilon))$$

and

$$\lim_{T\to\infty} T^{d+2-2\gamma-\varepsilon} \mathbf{E} \left(\frac{\lambda_i(x,T)}{\lambda_1^T} - p(0 \leadsto x, T) B_i \right)^2 = 0.$$

Consequently for any fixed $x \in \mathbb{Z}^d$ and $0 < \varepsilon < 1$

$$\lim_{T\to\infty} T^{1-\varepsilon} \left| \frac{\lambda_i(x,T)}{\lambda_1^T p(0 \leadsto x, T)} - B_i \right| = \lim_{T\to\infty} T^{1-\varepsilon} \left| \frac{1}{2} \left(\frac{2\pi T}{d} \right)^{d/2} \frac{\lambda_i(x,T)}{\lambda_1^T} - B_i \right| = 0 \quad a.s.$$

provided that $x \equiv T \pmod 2$.

9.6 Other cases

At the end of Section 9.2 we listed the cases which are interesting to study. In Sections 9.4 and 9.5 the results were formulated only in case $I(M) = 0$, $\det M \ne 0$, $\lambda_1 > 1$. This Section is devoted to study the other interesting cases.

Without having any new idea one can prove that Theorems 9.8, 9.9 and 9.10 remain true in the cases

(i) $I(M) = 0$, $\det M = 0$, $\lambda_1 = m_{00} + \mu m_{01} > 1$ without any change,

(ii) $I(M) = 2$, $m_{00} > 1$, $m_{01} > 0$ replacing λ_1 by m_{00} with $B_i = \lim_{t\to\infty} m_{00}^{-t} B(i,t)$,

(iii) $I(M) = 2$, $m_{11} > 1$, $m_{10} > 0$ replacing λ_1 by m_{11} with $B_i = \lim_{t\to\infty} m_{11}^{-t} B(i,t)$,

(iv) $I(M) = 1$, $m_{01} = 0$, $m_{00} > m_{11} > 1$ replacing λ_1 by m_{00} if $i = 0$ and by m_{11} if $i = 1$ with $B_0 = \lim_{t\to\infty} m_{00}^{-t} B(0,t)$ and $B_1 = \lim_{t\to\infty} m_{11}^{-t} B(1,t)$

(v) $I(M) = 1$, $m_{10} = 0$, $m_{11} > m_{00} > 1$ replacing λ_1 by m_{00} if $i = 0$ and by m_{11} if $i = 1$ with $B_0 = \lim_{t\to\infty} m_{00}^{-t} B(0,t)$ and $B_1 = \lim_{t\to\infty} m_{11}^{-t} B(1,t)$,

(vi) $I(M) = 1$, $m_{01} = 0$, $m_{11} > m_{00}$ and $m_{11} > 1$ replacing λ_1 by m_{11} with $B_i = \lim_{t\to\infty} m_{11}^{-t} B(i,t)$,

(vii) $I(M) = 1$, $m_{10} = 0$, $m_{00} > m_{11}$ and $m_{00} > 1$ replacing λ_1 by m_{00} with $B_i = \lim_{t\to\infty} m_{00}^{-t} B(i,t)$,

(viii) $I(M) = 1$, $m_{01} = 0$ and $m_{00} = m_{11} > 1$ replacing λ_1 by m_{11} if $i = 1$ and $m_{11}t^{1/t}$ if $i = 0$ with $B_1 = \lim_{t\to\infty} m_{11}^{-t} B(1,t)$ and $B_0 = \lim_{t\to\infty} m_{11}^{-t} t^{-1} B(0,t)$,

(ix) $I(M) = 1$, $m_{10} = 0$ and $m_{00} = m_{11} > 1$ replacing λ_1 by m_{11} if $i = 0$ and $m_{11}t^{1/t}$ if $i = 1$ with $B_0 = \lim_{t\to\infty} m_{11}^{-t} B(0,t)$ and $B_1 = \lim_{t\to\infty} m_{11}^{-t} t^{-1} B(1,t)$.

And there too, all order leaving,
The comet's image dread finds place:
But to the Lord's voice harkening
In order moves.

Madách: The Tragedy of Man
Translated by W. N. Loew

Part III
STRASSEN TYPE THEOREMS

Chapter 10

Infinitely many independent particles

10.1 The original form of Strassen's law

Let $\{W(t),\ t \geq 0\}$ be a Wiener process and let

$$w_T(x) = b_T W(xT) \qquad (0 \leq x \leq 1,\ T > 0)$$

where

$$b_T = (2T \log\log T)^{-1/2}$$

Let $S \subset C(0,1)$ be the set of those absolutely continuous functions (with respect to Lebesgue measure) for which

$$f(0) = 0 \qquad \text{and} \qquad \int_0^1 (f'(x))^2 dx \leq 1.$$

S is called *Strassen class*.

The set S is compact in $C(0,1)$ and this follows from

LEMMA 10.1 ([19] Lemma 1.3.1) *Let $f(\cdot)$ be a real valued function on $[0,1]$. The following two conditions are equivalent:*

(i) *f is absolutely continuous and $\int_0^1 (f'(x))^2 dx \leq 1$,*

(ii) $r \sum_{i=1}^{r} \left(f\left(\frac{i}{r}\right) - f\left(\frac{i-1}{r}\right)\right)^2 \leq 1$ *for any $r = 1, 2, \ldots$ and $f \in C(0,1)$.*

Now we formulate the celebrated Strassen's theorem as follows.

THEOREM 10.1 ([44] p.83) *The net $\{w_T(x),\ 0 \leq x \leq 1\}$ is relatively compact in $C(0,1)$ with probability 1 and the set of its limit points is S.*

The meaning of this statement is that there exists an event $\Omega_0 \subset \Omega$ of probability zero with the following two properties:

Property 1. For any $\omega \notin \Omega_0$ and any sequence of positive numbers $T_1 < T_2 < \ldots$ with $T_n \to \infty$ as $n \to \infty$ there exist a random subsequence $T_{i(j)}$ and a function $f \in S$ such that

$$w_{T_{i(j)}}(x) \to f(x) \quad \text{uniformly in} \quad [0,1].$$

In other words if T is big enough $w_T(x)$ can be approximated by a suitable element of S.

Property 2. For any $f \in S$ and $\omega \notin \Omega_0$ there exists a sequence of positive numbers $T_k = T_k(\omega, f)$ such that

$$w_{T_k}(x) \to f(x) \quad \text{uniformly in} \quad [0,1].$$

In other words any $f \in S$ can be approximated by $w_T(x)$ for some T big enough.

Now we turn to the d-dimensional generalization of Theorem 10.1.

Let $\{W^{(d)}(t) = (W_1(t), W_2(t), \dots, W_d(t)),\ t \geq 0\}$ be a Wiener process on $\mathbb{R}^d$ and consider

$$w_T^{(d)}(x) = b_T W^{(d)}(xT) \qquad (0 \leq x \leq 1,\ T > 0)$$

where

$$b_T = (2T \log\log T)^{-1/2}.$$

Let $S^d \subset (C(0,1))^d$ be the set of those $\mathbb{R}^d$-valued functions

$$f(x) = (f_1(x), f_2(x), \dots, f_d(x))$$

for which

(i) $f_i(x)$ is absolutely continuous with $f_i(0) = 0$,

(ii) $\int_0^1 \sum_{i=1}^{d} (f_i'(x))^2 dx \leq 1.$

Then we have

THEOREM 10.2 *The net $\{w_T^{(d)}(x),\ 0 \leq x \leq 1\}$ is relatively compact in $(C(0,1))^d$ with probability 1 and the set of its limit points is S^d.*

The proof of Theorem 10.2 can be easily obtained following the proof of Theorem 10.1 given in [19 p. 37]. Here we only sketch the main steps.

Now we recall that the proof of Theorem 10.1 in [19] based on Lemma 10.1 and on

LEMMA 10.2 ([19] Lemma 1.3.2) *Let r be a positive integer and $\alpha_1, \alpha_2, \dots, \alpha_r$ be a sequence of real numbers for which*

$$\sum_{i=1}^{r} \alpha_i^2 = 1.$$

Further let

$$S(T) = \alpha_1 W(T) + \alpha_2(W(2T) - W(T)) + \cdots + \alpha_r(W(rT) - W((r-1)T)).$$

Then

$$\limsup_{T\to\infty} b_T S(T) = -\liminf_{T\to\infty} b_T S(T) = 1 \quad a.s.$$

In order to prove Theorem 10.2 we give the following two lemmas.

LEMMA 10.3 *Let $f_1(\cdot), f_2(\cdot), \ldots, f_d(\cdot)$ be real valued functions on $[0,1]$. The following two conditions are equivalent:*

(i) *$f_1, f_2, \ldots, f_d$ are absolutely continuous and*

$$\sum_{i=1}^{d} \int_0^1 (f_i'(x))^2 \leq 1,$$

(ii)

$$r \sum_{j=1}^{d} \sum_{i=1}^{r} \left(f_j\left(\frac{i}{r}\right) - f_j\left(\frac{i-1}{r}\right)\right)^2 \leq 1 \quad \textit{for any} \quad r = 1, 2, \ldots$$

and $f_i \in C(0,1)$, $i = 1, 2, \ldots, d$.

Proof. It is a trivial consequence of Lemma 10.1.

LEMMA 10.4 *Let $W_1(t), W_2(t), \ldots, W_d(t)$ be independent Wiener processes, r be a positive integer and let α_{ij} $(i = 1, 2, \ldots, d,\ j = 1, 2, \ldots, r)$ be an array of real numbers for which*

$$\sum_{i=1}^{d} \sum_{j=1}^{r} \alpha_{ij}^2 = 1.$$

Further let

$$S(T) = \sum_{i=1}^{d} \sum_{j=1}^{r} \alpha_{ij}(W_i(jT) - W_i((j-1)T)).$$

Then

$$\limsup_{T\to\infty} b_T S(T) = -\liminf_{T\to\infty} b_T S(T) = 1 \qquad a.s.$$

Proof can be obtained repeating word by word, the proof of Lemma 10.2 given in [19 Lemma 1.3.2].

Now the **proof of Theorem 10.2** can be easily obtained by Finkelstein's method just like Theorem 10.1 was proved in [19 p. 39].

10.2 On the coverage of S

Let $\{W_i(t),\ t \geq 0\}$, $i = 1, 2, \ldots$ be a sequence of independent Wiener processes and let $\kappa(T) \geq 1$, $(T \geq 0)$ be a nondecreasing, integer valued function which takes all positive integer. Further let

$$w_T(i, x) = \beta_T W_i(xT) \quad (i = 1, 2, \ldots, \kappa(T),\ 0 \leq x \leq 1) \tag{10.1}$$

and

$$w_T = \{w_T(1,x), w_T(2,x), \ldots, w_T(\kappa(T),x),\ 0 \le x \le 1\}$$

where

$$\beta_T = (2T(\log\kappa(T) + \log\log T))^{-1/2}.$$

We are interested in the path behaviour of the random set $\{w_T\}$.

In order to formulate our results we introduce a few definitions.

Let $C_0(0,1)$ denote the set all continuous functions on $[0,1]$ with $f(0) = 0$ endowed with the sup-norm distance. Let F_T $(T > 0)$ be a family of subsets of $C_0(0,1)$. Let $F^{(1)}$ be the set of all limits of convergent sequences of the form $f_{T_n} \in F_{T_n}$ with $T_n \to \infty$, whereas $F^{(0)}$ is the set of limits as $T \to \infty$ of functions $f_T \in F_T$ whenever such limits exist. We will call

$$F^{(1)} = \limsup_{T\to\infty} F_T \qquad \text{and} \qquad F^{(0)} = \liminf_{T\to\infty} F_T.$$

In other words $f \in F^{(1)}$ if and only if there exists a sequence $T_n \to \infty$ such that for any n there exists an $f_{T_n} \in F_{T_n}$ for which $f_{T_n}(x) \to f(x)$ uniformly in x as $n \to \infty$. Whereas $f \in F^{(0)}$ if and only if for any T there exists an $f_T \in F_T$ such that $f_T(x) \to f(x)$ uniformly in x as $T \to \infty$.

Whenever $F^{(0)} = F^{(1)}$ this set will be called the limit of F_T denoting by

$$F^{(0)} = F^{(1)} = \lim_{T\to\infty} F_T.$$

For example if

$$F_{2T} = \{f(x) :\ f(x) = cx,\ -\infty < c < \infty\},$$
$$F_{2T+1} = \{f(x) :\ f(x) = cx^2,\ -\infty < c < \infty\}$$

$(T = 0, 1, 2, \ldots)$ then

$$\limsup_{T\to\infty} F_T = F_0 \cup F_1,$$
$$\liminf_{T\to\infty} F_T = \{f(x) :\ f(x) = 0\}.$$

Theorem 10.1 can be formulated as follows:

$$\limsup_{T\to\infty}\{w_T(x),\ 0 \le x \le 1\} = S$$

and

$$\liminf_{T\to\infty}\{w_T(x),\ 0 \le x \le 1\} = \emptyset.$$

Now we present two theorems on the limit behaviour of $\{w_T\}$.

THEOREM 10.3 ([23] Theorem 1.1) *Assume that*

$$\lim_{T\to\infty} \frac{\log \kappa(T)}{\log\log T} = c \in (0,\infty). \tag{10.2}$$

Then

$$\liminf_{T\to\infty} w_T = \left(\frac{c}{c+1}\right)^{1/2} \mathcal{S} \qquad a.s. \tag{10.3}$$

and

$$\limsup_{T\to\infty} w_T = \mathcal{S} \qquad a.s. \tag{10.4}$$

THEOREM 10.4 ([23] Theorem 1.2) *Assume that*

$$\lim_{T\to\infty} \frac{\log \kappa(T)}{\log\log T} = \infty. \tag{10.5}$$

Then

$$\lim_{T\to\infty} w_T = \mathcal{S} \qquad a.s. \tag{10.6}$$

The meaning of the above two theorems can be described as follows. Consider the set of the paths of our $\kappa(T)$ independent Wiener processes (transformed by (10.1)). This random set of $C(0,1)$ functions is denoted by w_T.

Then Theorem 10.4 tells us that assuming (10.5) we have

(i) for all T big enough and for all $f(\cdot) \in \mathcal{S}$ there exists a $0 < i = i(T,\omega) < \kappa(T)$ such that $w_T(i,x)$ $(0 \le x \le 1)$ will approximate the given $f(\cdot)$,

(ii) for all T big enough and for every $0 < i < \kappa(T)$ the function $w_T(i,x)$ $(0 \le x < 1)$ can be approximated by a suitable $f(\cdot) \in \mathcal{S}$.

Theorem 10.3 tells us that assuming (10.2) (i.e. we assume that $\kappa(T)$ is about $(\log T)^c$) we have

(i) for all T big enough and for all $f(\cdot) \in \left(\frac{c}{c+1}\right)^{1/2} \mathcal{S}$ there exists a $0 < i = i(T,\omega) < \kappa(T)$ such that $w_T(i,x)$ $(0 \le x \le 1)$ will approximate the given $f(\cdot)$,

(ii) for all T big enough and for every $0 < i < \kappa(T)$ the function $w_T(i,x)$ $(0 \le x \le 1)$ can be approximated by a suitable $f(\cdot) \in \mathcal{S}$.

Very likely the d-dimensional generalizations of Theorems 10.3 and 10.4 hold true. Here we only prove that the d-dimensional version of (10.6) holds assuming (10.5).

THEOREM 10.5 *Let*

$$W_i(t) = (W_{i1}(t), W_{i2}(t), \ldots, W_{id}(t)) \in \mathbb{R}^d \qquad (i = 1, 2, \ldots)$$

be a sequence of independent Wiener processes and let $\kappa(T) \geq 1$ *be a nondecreasing, integer valued function satisfying* (10.5). *Further let*

$$w_T(i, x) = \beta_T W_i(xT), \qquad (i = 1, 2, \ldots, \kappa(T),\ 0 \leq x \leq 1)$$

and

$$w_T = \{w_T(1, x), w_T(2, x), \ldots, w_T(\kappa(T), x),\ 0 \leq x \leq 1\}$$

where

$$\beta_T = (2T(\log \kappa(T) + \log \log T))^{-1/2} \sim (2T \log \kappa(T))^{-1/2}.$$

Then

$$\lim_{T \to \infty} w_T \in \mathcal{S}^d \qquad a.s.$$

Before giving the proof we introduce a few notations and present a few lemmas. Let

$$\nu = \nu(T, M, (j_1, j_2, \ldots, j_d)) = \#\{i:\ 1 \leq i \leq \kappa(T),\ w_T(i, 1) \in \mathcal{B}\}$$

where

$$\mathcal{B} = \mathcal{B}(j_1, j_2, \ldots, j_d) = \left\{(y_1, y_2, \ldots, y_d):\ \frac{j_\ell}{M} \leq y_\ell < \frac{j_\ell + 1}{M},\ \ell = 1, 2, \ldots, d\right\},$$

and $T > 0,\ M = 1, 2, \ldots,\ j_\ell = 0, \pm 1, \pm 2, \ldots,\ \ell = 1, 2, \ldots, d.$
Then we have

LEMMA 10.5

$$\mathbf{P}\{\nu = k\} = \binom{\kappa(T)}{k} p^k (1 - p)^{\kappa(T) - k} \qquad (k = 0, 1, 2, \ldots, \kappa(T)) \tag{10.7}$$

where

$$\begin{aligned} p &= p(\kappa(T), j) = p(\kappa(T), j_1, j_2, \ldots, j_d) = \\ &= \left(\frac{\log \kappa(T)}{\pi}\right)^{d/2} \prod_{\ell=1}^{d} \int_{j_\ell/M}^{(j_\ell+1)/M} \exp(-u^2 \log \kappa(T)) du = \\ &= M^{-d} \left(\frac{\log \kappa(T)}{\pi}\right)^{d/2} \kappa(T)^{-(\varsigma_1^2 + \varsigma_2^2 + \cdots + \varsigma_d^2) M^{-2}}, \\ j &= (j_1, j_2, \ldots, j_d). \end{aligned}$$

and

$$j_\ell < \varsigma_\ell < j_\ell + 1 \ (\ell = 1, 2, \ldots, d).$$

Hence

$$\mathbf{E}\nu = M^{-d}\left(\frac{\log \kappa(T)}{\pi}\right)^{d/2} \kappa(T)^{1-(\varsigma_1^2+\varsigma_2^2+\cdots+\varsigma_d^2)M^{-2}},$$
$$\operatorname{Var}\nu \sim \mathbf{E}\nu \qquad \textit{as} \qquad T \to \infty.$$

Further for any $K > 0$ and T big enough

$$\mathbf{P}\{|\nu - \mathbf{E}\nu| \geq (3K\mathbf{E}\nu \log T)^{1/2}\} \leq T^{-K}$$

provided that

$$\sum_{\ell=1}^{d} (j_\ell + 1)^2 < M^2.$$

Consequently for any $\varepsilon > 0$, $K > 0$ and T big enough

$$\mathbf{P}\{\nu < \mathbf{E}\nu - (\mathbf{E}\nu)^{1/2+\varepsilon}\} \leq T^{-K}$$

if

$$\sum_{\ell=1}^{d} (j_\ell + 1)^2 < M^2.$$

Proof is trivial.

LEMMA 10.6 *Let*

$$\sum_{\ell=1}^{d} j_\ell^2 > M^2(1+\varepsilon).$$

Then

$$\mathbf{P}\{\nu \geq 1\} = 1 - (1-p)^{\kappa(T)} \leq (\kappa(T))^{-\varepsilon/2}$$

if T is big enough.

Proof. Lemma 10.6 is a trivial consequence of (10.7).

Introduce the following further notations

$$\begin{aligned}
j^{(1)} &= (j_{11}, j_{12}, \ldots, j_{1d}),\\
j^{(2)} &= (j_{21}, j_{22}, \ldots, j_{2d}),\\
\ldots \quad & \ldots\ldots\ldots\\
j^{(r)} &= (j_{r1}, j_{r2}, \ldots, j_{rd}),
\end{aligned}$$

$$J = \begin{pmatrix} j^{(1)} \\ j^{(2)} \\ \ldots \\ j^{(r)} \end{pmatrix} = \begin{pmatrix} j_{11}, & j_{12}, & \ldots, & j_{1d} \\ j_{21}, & j_{22}, & \ldots, & j_{2d} \\ \ldots & \ldots & \ldots & \ldots \\ j_{r1}, & j_{r2}, & \ldots, & j_{rd} \end{pmatrix},$$

$$\|j^{(\alpha)}\|^2 = \sum_{\beta=1}^{d} j_{\alpha\beta}^2,$$

$$\|J\|^2 = \sum_{\alpha=1}^{r} \|j^{(\alpha)}\|^2 = \sum_{\beta=1}^{d}\sum_{\alpha=1}^{r} j_{\alpha\beta}^2,$$

$$V_\alpha = V_\alpha(T,M,r,J) =$$

$$= \#\left\{i:\ 1 \le i \le \kappa(T),\ \bigcap_{\mu=1}^{\alpha}\{w_T(i,\mu) - w_T(i,\mu-1) \in \mathcal{B}_\mu\}\right\}$$

where

$$\mathcal{B}_\mu = \mathcal{B}_\mu(j_{\mu 1}, j_{\mu 2}, \ldots, j_{\mu d}) = \left\{(y_1, y_2, \ldots, y_d):\ \frac{j_{\mu\ell}}{M} \le y_\ell < \frac{j_{\mu\ell}+1}{M},\ \ell = 1,2,\ldots,d\right\}$$

and $\alpha = 1,2,\ldots,r$.

Note that by Lemma 10.5 for any $K > 0$, $\varepsilon > 0$ and T big enough

$$\mathbf{P}\{V_1 < \mathbf{E}V_1 - (\mathbf{E}V_1)^{1/2+\varepsilon}\} \le T^{-K}$$

provided that

$$\sum_{\beta=1}^{d}(j_{1\beta}+1)^2 < M^2$$

where

$$\mathbf{E}V_1 = M^{-d}\left(\frac{\log\kappa(T)}{\pi}\right)^{d/2}\kappa(T)^{1-(\varsigma_{11}^2+\varsigma_{12}^2+\cdots+\varsigma_{1d}^2)M^{-2}},$$

$$j_{1\beta} < \varsigma_{1\beta} < j_{1\beta}+1 \qquad (\beta = 1,2,\ldots,d).$$

Similarly

$$\mathbf{P}\{V_2 < \mathbf{E}V_2 - (\mathbf{E}V_2)^{1/2+\varepsilon}\} \le T^{-K}$$

where

$$\mathbf{E}V_2 = \kappa(T)p(\kappa(T), j^{(1)})p(\kappa(T), j^{(2)}) \sim M^{-2d}\left(\frac{\log\kappa(T)}{\pi}\right)^{d}\kappa(T)^{Q},$$

$$Q = 1 - M^{-2}\sum_{i=1}^{2}\sum_{\ell=1}^{d}\varsigma_{i\ell}^2$$

and

$$\frac{j_{i\ell}}{M} \le \varsigma_{i\ell} < \frac{j_{i\ell}+1}{M}.$$

Continuing the procedure we get

LEMMA 10.7 *Assume that*

$$\bar{j}^2 = \sum_{\beta=1}^{d} \sum_{\alpha=1}^{r} (j_{\alpha\beta} + 1)^2 < M^2.$$

Then for any $\varepsilon > 0$, $K > 0$, $r = 1, 2, \ldots$ and T big enough we have

$$\mathbf{P}\left\{ V_r < (1-\varepsilon)\left(M^{-d} \left(\frac{\log \kappa(T)}{\pi} \right)^{d/2} \right)^r \kappa(T)^{1-\bar{j}^2 M^{-2}} \right\} \le T^{-K}.$$

Similarly by Lemma 10.6 we get

LEMMA 10.8 *Assume that*

$$\bar{j}^2 > (1+\varepsilon)M^2.$$

Then

$$\mathbf{P}\{V_r \ge 1\} \le \kappa(T)^{-\varepsilon/2}.$$

LEMMA 10.9 *For $0 < c < 1$ we have*

$$\limsup_{T\to\infty} \sup_{1 \le i \le \kappa(T)} \sup_{0 \le t \le T - cT} \frac{W_i(t+cT) - W_i(t)}{(2T \log \kappa(T))^{1/2}} \le c^{1/2} \quad a.s.$$

Proof can be obtained by routine methods. (See the proof of Theorem 1.2.1 in [19]).

Proof of Theorem 10.5. Lemmas 10.3, 10.7 and 10.9 imply

$$\liminf_{T\to\infty} w_T \supset S^d \qquad \text{a.s.}$$

Similarly Lemmas 10.3 and 10.8 imply

$$\limsup_{T\to\infty} w_T \subset S^d \qquad \text{a.s.}$$

Hence we have Theorem 10.5.

Chapter 11

Branching random walk

11.1 The question

A sequence $\{x_s, x_{s+1}, \ldots, x_t\}$ ($x_i \in \mathbb{Z}^d$, $0 \leq s \leq i \leq t$) is called a *nearest-neighbour* (s,t)*-sequence* if x_{s+i} and x_{s+i+1} ($i = 0, 1, 2, \ldots, t-s-1$) are neighbours in $\mathbb{Z}^d$ i.e. $|x_{s+i+1} - x_{s+i}| = 1$. Consider a branching random walk on $\mathbb{Z}^d$ and let $\{x_s, x_{s+1}\}$ be a nearest-neighbour $(s, s+1)$-sequence. It is called an $(s,s+1)$*-path* of the branching random walk if there exists a particle located in x_s at time s which has at least one offspring located in x_{s+1} at time $s+1$. We say that *a particle belongs to* the $(s,s+1)$-path $\{x_s, x_{s+1}\}$ if either it is located in x_s at time s and it has an offspring located in x_{s+1} at time $s+1$ or it is located in x_{s+1} at time $s+1$ and it has an ancestor located in x_s at time s.

By induction, a nearest-neighbour $(s,t+1)$-sequence $\{x_s, x_{s+1}, \ldots, x_{t+1}\}$ is called an $(s,t+1)$-path if $\{x_s, x_{s+1}, \ldots, x_t\}$ is an (s,t)-path and there is a particle located in x_t at time t belonging to the (s,t)-path $\{x_s, x_{s+1}, \ldots, x_t\}$ which has an offspring located in x_{t+1} at time $t+1$. A particle located in x_{t+1} at time $t+1$ is belonging to the $(s,t+1)$-path $\{x_s, x_{s+1}, \ldots, x_{t+1}\}$ if it has an ancestor located in x_t at time t, belonging to the (s,t)-path $\{x_s, x_{s+1}, \ldots, x_t\}$.

Consider all $(0,T)$-paths of the underlying branching random walk. Clearly, the number of these paths is $B(T)$. Our goal is to characterize the set of these paths. It turns out that the properties of these paths depend strongly on m. We consider the case $m > 2d$ only. The general case, for an arbitrary $m > 1$ will be treated for branching Wiener process.

11.2 The case $m > 2d$

Our main result is

THEOREM 11.1 *On the set $\{B > 0\}$ there exists a r.v. $N = N(\omega)$ taking even positive numbers only such that any $(N, N+T)$-nearerst-neighbour sequence*

$$\{x_N = 0, x_{N+1}, \ldots, x_{N+T}\} \qquad (T = 1, 2, \ldots)$$

is an $(N, N+T)$-path of the branching random walk.

Proof. Let $\{x_s, x_{s+1}, \dots, x_t\}$ be a nearest-neighbour sequence in $\mathbb{Z}^d$ and let

$$\tilde{\lambda}(s,t) = \tilde{\lambda}(x_s, x_{s+1}, \dots, x_t\}$$

be the number of the (s,t)-paths $(x_s, x_{s+1}, \dots, x_t)$ i.e. $\tilde{\lambda}(s,t)$ is the number of those particles which are located in x_t at time t and belonging to the (s,t)-path $(x_s, x_{s+1}, \dots, x_t)$. Naturaly we define

$$\tilde{\lambda}(s) = \tilde{\lambda}(s,s) = \lambda(0,s).$$

Observe that for any $0 < \varepsilon < 1$ and for any nearest-neighbour $(s,t+1)$-sequence $\{x_s, x_{s+1}, \dots, x_{t+1}\}$

$$\mathbf{P}\left\{\tilde{\lambda}(s,t+1) < (1-\varepsilon)\frac{m}{2d}\tilde{\lambda}(s,t) \;\middle|\; \tilde{\lambda}(s,t)\right\} \le \exp(-O(1)\varepsilon^2\tilde{\lambda}(s,t)).$$

Consequently

$$\mathbf{P}\left\{\tilde{\lambda}(s,t+1) < (1-\varepsilon)\frac{m}{2d}K \;\middle|\; \tilde{\lambda}(s,t) \ge K\right\} \le \exp(-O(\varepsilon^2 K)).$$

Similarly if x_{t+2} is a neightbour of x_{t+1} then

$$\mathbf{P}\left\{\tilde{\lambda}(s,t+2) \ge \left((1-\varepsilon)\frac{m}{2d}\right)^2 K \;\middle|\; \tilde{\lambda}(s,t) \ge (1-\varepsilon)\frac{m}{2d}K\right\} \ge$$

$$\ge 1 - \exp\left(-O\left((1-\varepsilon)\varepsilon^2\frac{m}{2d}K\right)\right).$$

Hence

$$\mathbf{P}\left\{\tilde{\lambda}(s,t+2) \ge \left((1-\varepsilon)\frac{m}{2d}\right)^2 K,\, \tilde{\lambda}(s,t+1) \ge (1-\varepsilon)\frac{m}{2d}K \;\middle|\; \tilde{\lambda}(s,t) \ge K\right\} =$$

$$= \mathbf{P}\left\{\tilde{\lambda}(s,t+2) \ge \left((1-\varepsilon)\frac{m}{2d}\right)^2 K \;\middle|\; \tilde{\lambda}(s,t+1) \ge (1-\varepsilon)\frac{m}{2d}K,\, \tilde{\lambda}(s,t) \ge K\right\} \times$$

$$\times \mathbf{P}\left\{\tilde{\lambda}(s,t+1) \ge (1-\varepsilon)\frac{m}{2d}K \;\middle|\; \tilde{\lambda}(s,t) \ge K\right\} \ge$$

$$\ge (1 - \exp(-O(\varepsilon^2 K)))\left(1 - \exp\left(-O\left((1-\varepsilon)\varepsilon^2\frac{m}{2d}K\right)\right)\right).$$

By induction we have

$$\mathbf{P}\left\{\prod_{i=1}^{n}\left\{\tilde{\lambda}(s,t+i) \ge \left((1-\varepsilon)\frac{m}{2d}\right)^i K\right\} \;\middle|\; \tilde{\lambda}(s,t) \ge K\right\} \ge$$

$$\prod_{i=1}^{n}\left(1 - \exp\left(-O\left(\varepsilon^2 K\left((1-\varepsilon)\frac{m}{2d}\right)^{i-1}\right)\right)\right) \ge 1 - \exp(-O(\varepsilon^2 K)).$$

In case $t = s = 2u$ we get

$$\mathbf{P}\left\{\prod_{i=1}^{n}\left\{\tilde{\lambda}(2u,2u+i) \geq \left((1-\varepsilon)\frac{m}{2d}\right)^{i} K\right\} \,\middle|\, \lambda(0,2u) \geq K\right\} \leq 1 - \exp(-O(\varepsilon^2 K)).$$

Since by (4.31)

$$\lambda(0,2u) \geq m^{2u} u^{-d/2-\varepsilon} \qquad \text{a.s.}$$

for any $\varepsilon > 0$ and for all but finitely many u, we have Theorem 11.1.

A simple consequence of the above theorem is the following

THEOREM 11.2 *On the set $\{B > 0\}$ there exists a r.v. $N = N(\omega)$ taking even positive numbers such that for any x with $|x| < 2T - N$ and with $x \equiv 0 \pmod 2$ we have*

$$\lambda(x, 2T) > 0$$

and

$$\lambda(x, 2T) = 0 \quad \text{if} \quad |x| > 2T.$$

11.3 Branching Wiener process

Let $W_1(t), W_2(t), \ldots, W_{B(T)}(t)$ $(0 \leq t \leq T)$ be the $(0,T)$-branches of a branching Wiener process in $\mathbb{R}^d$. Clearly $W_i(\cdot)$ $(i = 1,2,\ldots,B(T))$ are non-independent Wiener processes. Further let

$$w_T(i,x) = T^{-1} W_i(xT) \qquad (i = 1,2,\ldots,B(T),\ 0 \leq x \leq 1)$$

and

$$w_T = \{w_T(1,x), w_T(2,x), \ldots, w_T(B(T),x) \quad 0 \leq x \leq 1\}$$

be the set of the normalized $(0,T)$-branches of the underlying branching Wiener process.

Let $Z^d \subset (C(0,1))^d$ be the set of those $\mathbb{R}^d$-valued functions

$$f(x) = (f_1(x), f_2(x), \ldots, f_d(x))$$

for which

(i) $f_i(x)$ are absolutely continuous on $(0,1)$ and $f_i(0) = 0$ $(i = 1,2,\ldots,d)$,

(ii) $\displaystyle\int_0^{\lambda} \sum_{i=1}^{d} (f_i'(x))^2 dx \leq \lambda \qquad (0 \leq \lambda \leq 1).$

Note that $Z^d \subset S^d$ and it is compact in $(C(0,1))^d$. This follows from the following:

LEMMA 11.1 *Let $f_1(\cdot), f_2(\cdot), \ldots, f_d(\cdot)$ be real-valued functions on $[0,1]$. Then the following two conditions are equivalent*

(i) $f_1, f_2, \ldots, f_d$ *are absolutely continuous and for any* $0 \le \lambda \le 1$

$$\int_0^\lambda \sum_{i=1}^d (f_i'(x))^2 dx \le \lambda,$$

(ii)

$$r \sum_{j=1}^d \sum_{i=1}^{[\lambda r]} \left(f_j\left(\frac{i}{r}\right) - f_j\left(\frac{i-1}{r}\right)\right)^2 \le \lambda$$

for any $r = 1, 2, \ldots$ *and for any* $0 \le \lambda \le 1$.

Proof. Since for any $0 \le a < b \le 1$ and $j = 1, 2, \ldots, d$ by the Cauchy inequality

$$\frac{(f_j(b) - f_j(a))^2}{b-a} \le \int_a^b (f_j'(x))^2 dx,$$

(i) implies (ii). Since it is assumed that the inequality of (ii) holds for any r, we have (i) as a consequence of (ii).

We are interested in the limit behaviour of the random set w_T as $T \to \infty$ and present the following:

THEOREM 11.3 *On the set $\{B > 0\}$ we have*

$$\lim_{T\to\infty} w_T = (2 \log m)^{1/2} \mathcal{Z}^d \qquad a.s.$$

At first we introduce some notations. Let

$$\begin{aligned}
j_{\alpha\beta} &= 0, \pm 1, \pm 2, \ldots, \quad \alpha = 1, 2, \ldots, r, \quad \beta = 1, 2, \ldots, d,\\
j^{(1)} &= (j_{11}, j_{12}, \ldots, j_{1d}),\\
j^{(2)} &= (j_{21}, j_{22}, \ldots, j_{2d}),\\
\ldots & \quad \ldots\ldots\\
j^{(r)} &= (j_{r1}, j_{r2}, \ldots, j_{rd}),\\
J(\mu) &= \begin{pmatrix} j^{(1)} \\ j^{(2)} \\ \ldots \\ j^{(\mu)} \end{pmatrix} = \begin{pmatrix} j_{11}, & j_{12}, & \ldots, & j_{1d} \\ j_{21}, & j_{22}, & \ldots, & j_{2d} \\ \ldots & \ldots & \ldots & \ldots \\ j_{\mu 1}, & j_{\mu 2}, & \ldots, & j_{\mu d} \end{pmatrix}, \quad (\mu = 2, 3, \ldots, r)\\
\|j^{(\alpha)}\|^2 &= \sum_{\beta=1}^d j_{\alpha\beta}^2,\\
\|J(\mu)\|^2 &= \sum_{\alpha=1}^\mu \|j^{(\alpha)}\|^2 = \sum_{\beta=1}^d \sum_{\alpha=1}^\mu j_{\alpha\beta}^2.
\end{aligned}$$

Let

$$W_1^{(\mu)}(t), W_2^{(\mu)}(t), \ldots, W_{B(\mu T)}^{(\mu)}(t) \qquad (0 \le t \le \mu T,\ \mu = 2,3,\ldots,r)$$

be the $(0, \mu T)$–branches of the underlying branching Wiener process and let

$$\begin{aligned} V_\mu &= V_\mu(j_2, j_3, \ldots, j_\mu) = \\ &= \#\{i:\ T^{-1}(W_i^{(\mu)}(\ell T) - W_i^{(\mu)}((\ell-1)T)) \in \mathcal{B}_\ell,\ \ell = 2,3,\ldots,\mu\} \end{aligned}$$

where $\mu = 2,3,\ldots,r$ and

$$\mathcal{B}_\ell = \mathcal{B}_\ell(j_\ell) = \left\{(y_1, y_2, \ldots, y_d):\ \frac{j_{\ell\beta}}{M} \le y_\beta < \frac{j_{\ell\beta}+1}{M},\ \beta = 1,2\ldots,d\right\}$$

for any $\ell = 1,2,\ldots,\mu$.

Finally let

$$V_{01} = V_{01}(j_1) = \#\{i:\ i = 0,1,\ldots,B(T),\ T^{-1}W_i(T) \in \mathcal{B}_1(j_1)\}$$

and

$$\begin{aligned} V_{0\mu} &= V_{0\mu}(j_1, j_2, \ldots, j_\mu) = \\ &= \{i:\ T^{-1}(W_i^{(\mu)}(\ell T) - W_i^{(\mu)}((\ell-1)T)) \in \mathcal{B}_\ell,\ \ell = 1,2,\ldots,\mu\}. \end{aligned}$$

Now we present a few lemmas.

LEMMA 11.2 *Let $W_1(\cdot)$ and $W_2(\cdot)$ be two independent Wiener processes on $\mathbb{R}^d$ and define*

$$W_s(t) = \begin{cases} W_1(t) & \text{if } t \le s, \\ W_1(s) + W_2(t-s) & \text{if } t \ge s. \end{cases}$$

Then for any Borel set $A \subset \mathbb{R}^d$ and for any $0 \le s \le t < \infty$ we have

$$\mathbf{P}\{W_s(t) \in A \mid W_1(t) \in A\} \ge \mathbf{P}\{W_1(t) \in A\}.$$

Proof is trivial.

LEMMA 11.3 *For any $0 < \varepsilon < 1$ we have*

$$\mathbf{P}\left\{V_2(j_2) < \frac{1-\varepsilon}{(2\pi)^{d/2}} \frac{m^{2T}}{M^d} B \exp\left(-\frac{x_2^2}{2}T\right)\right\} \le e^{-CT}$$

if

$$x_2^2 = M^{-2} \sum_{\beta=1}^{d} (j_{2\beta}+1)^2 < 2\log m$$

and C is a small enough positive constant.

Proof. $B(T) \sim m^T B$ particles are living at time T. Such a particle has an increment during the time interval $[T, 2T]$ belonging to $\mathcal{B}_2(j_2)$ with probability larger than

$$(2\pi)^{-d/2} M^{-d} \exp\left(-\frac{x_2^2}{2} T\right).$$

Hence we get Lemma 11.3 by Lemma 11.2.

LEMMA 11.4

$$\mathbf{P}\{V_1(j_1) \neq 0\} \leq e^{-CT}$$

if

$$M^{-2} \sum_{\beta=1}^{d} j_{1\beta}^2 > 2 \log m.$$

Proof is trivial.

The above two lemmas can be easily generalized as follows:

LEMMA 11.5 *For any $0 < \varepsilon < 1$ and $\mu = 2, 3, \ldots, r$ we have*

$$\mathbf{P}\left\{V_\mu(j_2, j_2, \ldots, j_\mu) < \left(\frac{1-\varepsilon}{(2\pi)^{d/2}} \frac{1}{M^d}\right)^{\mu-1} m^{\mu T} B \exp\left(-\frac{T}{2} \sum_{i=2}^{\mu} x_i^2\right)\right\} \leq e^{-CT}$$

if

$$\sum_{i=2}^{\mu} x_i^2 = M^{-2} \sum_{i=2}^{\mu} \sum_{\beta=1}^{d} (j_{i\beta} + 1)^2 < 2 \log m$$

and C is a small enough positive constant.

LEMMA 11.6 *For any $0 < \varepsilon < 1$ and $\mu = 1, 2, \ldots, r$ we have*

$$\mathbf{P}\{V_{0\mu}(j_1, j_2, \ldots, j_\mu) \neq 0\} \leq e^{-CT}$$

if

$$M^{-2} \sum_{i=1}^{\mu} \sum_{\beta=1}^{d} j_{i\beta}^2 > 2 \log m.$$

Proof of Theorem 11.3. Having Lemmas 11.5 and 11.6 and repeating the proof of Theorem 10.5 we have Theorem 11.3.

We say that the $\{w_T(i, T),\ i = 1, 2, \ldots, B(T)\}$ ε–covers the ball $C(0, r) \subset \mathbb{R}^d$ if for any $\xi \in C(0, r)$ there exists an i $(1 \leq i \leq B(T))$ such that $\|\xi - w_T(i, T)\| \leq \varepsilon$. Now we present the following trivial consequence of Theorem 11.3.

THEOREM 11.4 *On the set $\{B > 0\}$ for any $\varepsilon > 0$ there exists a $T_0 = T_0(\varepsilon, \omega)$ such that $\{w_T(i, T),\ i = 1, 2, \ldots, B(T)\}$ ε-covers the ball $C(0, (2 \log m)^{1/2})$ if $T > T_0$. Further for any $i = 1, 2, \ldots, B(T)$ $w_T(i, T) \in C(0, (2 \log m)^{1/2} + \varepsilon)$.*

Theorem 11.4 tells us that for any $\xi \in \mathcal{C}(0,(2\log m)^{1/2}T)$ and for any $\varepsilon > 0$ there exists an i $(i = 1,2,\ldots,B(T))$ such that

$$W_i(T) \in \mathcal{C}(\xi, \varepsilon T)$$

if T is big enough. It is natural to ask if we could find an i $(i = 1,2,\ldots,B(T))$ such that

$$W_i(T) \in \mathcal{C}(\xi, \varepsilon).$$

It turns out that it is so, even it is true if ε is replaced by $\varepsilon = \varepsilon_T = e^{-\lambda T}$ provided that λ is small enough. In fact we have

THEOREM 11.5 *On the set $\{B > 0\}$ for any $0 < \delta < \log m$ and for any $\xi \in \mathcal{C}(0,(2\log m - 2\delta)^{1/2}T)$ there exists a $T_0 = T_0(\delta,\omega)$ and an $i = i(\xi,T,\delta,\omega)$ $(i = 1,2,\ldots,B(T))$ such that*

$$W_i(T) \in \mathcal{C}(\xi, \varepsilon_T)$$

if $T \geq T_0$ and

$$\varepsilon_T = \exp\left(-\frac{\delta}{2d}T\right).$$

Proof. Follow the proof of Theorem 11.3 with

$$M = \exp\left(\frac{\delta}{2d}T\right).$$

Clearly Theorem 11.5 implies (6.27).

Historical overview

Part I

It was said in the Introduction that in the treatment of the random motion of a random field we are free to choose the law that governs the initial random field and the law governing the motion of the particles.

In the literature mostly the following two initial random fields are investigated:

(i) Poisson field of parameter λ on $\mathbb{R}^d$,

(ii) the field described by an array ε_x $(x \in \mathbb{Z}^d)$ of i.i.d.r.v's puting ε_x particles in x.

As far as the motions are concerned the following ones are investigated:

(a) independent Wiener processes in $\mathbb{R}^d$ starting from each point of the initial field, these motions are considered in case (i),

(b) in case (ii) mostly independent, identically distributed, $\mathbb{Z}^d$ valued random walks are considered starting also from each point of the initial field.

In case (b) occasionally the holding times (the times needed for one step) are i.i.d. positive valued r.v.'s (mostly exponential) with expectation 1.

In case (i)–(a) frequently the case $\lambda \to \infty$ is treated. The first paper of this type is due to Walsh [55]. He investigates the process $s(A,t)$ whenever $A \subset \mathbb{R}^d$ is a Borel set and $t \geq 0$. In fact, instead of the process $s(A,t)$ the process

$$\eta_t^\lambda(\phi) = \lambda^{-1/2} \sum_{i=1}^{\infty} \phi(P_i(t))$$

is investigated where $\phi(\cdot)$ is in the Schwartz space and $\lambda \to \infty$. Walsh proves that in limit $\eta_t^\lambda(\phi)$ satisfies a stochastic partial differential equation and gives a number of properties of the solution of this equation. The solution of this equation is called by Walsh *Brownian density process.*

Other important papers like [1] and [2] by Adler et al., investigate the intersection local times of the Brownian density process. In the latter two papers the case of charged particles i.e. the process

$$(\eta_t^\lambda(\phi))^{+,-} = \lambda^{-1/2} \sum_{i=1}^{\infty} \sigma_i \phi(P_i(t))$$

is also treated, where $\sigma_1, \sigma_2, \ldots$ is a sequence of i.i.d.r.v.'s independent from $P(t)$ with

$$\mathbf{P}\{\sigma_1 = 1\} = \mathbf{P}\{\sigma_1 = -1\} = 1/2.$$

The case (ii)–(b) is treated by Port [40, 41]. He proves the strong law of large numbers and the central limit theorem for $N(t)$ and $D(t)$ when ε_x's are i.i.d. Poisson r.v.'s. Weiss [56, 57] considered the same problem assuming weaker conditions on ε_x's. Having even more general conditions on the field $\{\varepsilon_x\}$ and the random walks the corresponding results are proved by Port et al. in [42]. The case when the holding times are i.i.d.r.v.'s is treated by Cox and Griffeath in [17] where also large deviation results are given for $N(t)$ and $D(t)$.

The material of Chapter 2 follows the papers [46], [47]. It is worth while to mention that the properties of $S(T)$ are very similar to the properties of

$$\sup_{0 \le t \le T} (\pi(t+1) - \pi(t)) = m(T)$$

where $\pi(t)$ is a Poisson process on $\mathbb{R}^1$ (cf. [8,9] by Auer et al. and [49] by Révész). However, the results proved for $m(T)$ are sharper than the corresponding results for $S(T)$. It is natural to ask whether the exactness obtained for $m(T)$ can be achieved for $S(T)$ as well.

Part II

The theory of branching processes is a very rich and developing area of probability theory. Here we mention only books by Asmussen and Hering [3], Athreya and Ney [6] resp. Harris [28] where the reader can find the most important information on the subject. In Sections 4.1 and 7.1 only those simplest results are given which are used in the sequel. For example, the condition that the variance $\sigma^2 = \operatorname{Var} Y(i,j) < \infty$ is superfluous for most purposes. In fact Kesten and Stigum [30, 31] proved that the first part of (4.2) is true if and only if

$$\mathbf{E}Y(i,j) \log Y(i,j) < \infty.$$

A striking question is the distribution of B (cf. Theorem 4.2). The best results are due to Biggins and Bingham [14].

The multitype branching process is also a popular subject of the probability theory. Here we mention only the nice survey [37] by Mode on the subject. In fact the results and proofs of Section 9.1 are nearly the same as the corresponding part of [37]. Moreover in [37] not only the two-type but the general multitype branching process is also treated.

Branching random walks resp. branching diffusion processes were first considered by Moyal [38] and Skorohod [50]. One of the first results is due to Athreya and Ney [7] where the mean of branching random walk is treated in super- as well as in subcritical case. On the history of branching random walks we mention the survey [39] of Ney.

The results of Sections 4.4 and 4.5 are due to Révész [48]. They are closely related to the papers [4, 5] by Asmussen and Kaplan. In fact Asmussen and Kaplan consider much more general random walks than those of Sections 4.4 and 4.5. Otherwise they restrict themselves to the one dimensional case ($d = 1$) and they do not study the rate of convergence. Hopefully the idea given in Section 6.6 gives help to treat more general random walks as well. A d-dimensional extension of some of the results of [4, 5] is given in [54] by K. Uchiyama. Asmussen and Kaplan consider the case when the lifetime is 1 (cf. Section 4.2), while Uchiyama consider exponential lifetime (cf. Section 4.7).

The covariances of $\lambda(x,t)$ and $\Lambda(x,t)$ are treated by Sawyer [49].

The multitype branching random walk in case $d = 1$ is treated by Bramson et al. in [15].

Perhaps the biggest part of the literature on branching random walks investigates the critical case ($m = 1$). Most of the papers consider the case when

(i) the initial random field is Poisson in $\mathbb{R}^d$ and $\lambda \to \infty$,

(ii) the motions are independent $\mathbb{R}^d$ valued Wiener processes,

(iii) the lifetimes are independent exponentially distributed r.v.'s.

In the already mentioned paper [55] Walsh also investigates this case. The most complete survey is due to Dawson [20]. He (and many others) consider the branching random walk as a sequence of measures on $\mathbb{R}^d$. (The measure of a domain, at time t, is the number of particles in the domain at time t.) Hence Dawson investigates the properties of a measure valued process and as a special case he gets the branching random walk.

Here we wish to discuss two further papers. The first one is [26] by J. Fleischman (see also [21] by D. Dawson and G. Ivanoff). Here the "critical branching random walk starting with one particle" is treated in the case when the motions are governed by independent Wiener processes and the lifetime is random. The main results, in our language, are

$$\lim_{t\to\infty} \mathbf{P}\{\lambda(0,2t) < xt^{1/2} \mid B(t) > 0\} = F(x) \qquad (x \geq 0)$$

where $d = 1$ and $F(\cdot)$ is an unknown distribution function and

$$\lim_{t\to\infty} C_1 \log t \mathbf{P}\{\lambda(0,2t) > C_2 x t^{1/2} \mid B(t) > 0\} = e^{-x}$$

where $d = 2$ and C_1, C_2 are positive constants. Note that these theorems are not proved in case of the model treated above but they are very likely true.

The second paper to be mentioned is [18] by Cox and Griffeath where the "critical branching random walks of a random field" is treated. Here the initial field is Poisson, the motion is Wiener and the lifetime is exponential. Again formulating the results in our language

$$\lim_{T\to\infty} \mathbf{P}\left\{\frac{L(0,T)-T}{\sigma b_T} < x\right\} = \Phi(x)$$

where

$$L(0,T) = \sum_{t=0}^{T} \Lambda(0,t),$$

$$b_T = \begin{cases} T^{3/4} & \text{if } d = 3, \\ (T\log T)^{1/2} & \text{if } d = 4, \\ T^{1/2} & \text{if } d \geq 5 \end{cases}$$

are given constants depending on d. In case $d = 2$ the sequence $T^{-1}L(0,T)$ weakly converges to a nontrivial infinitely divisible r.v. We have to emphasize again that the above results are not proved in case of our model. We mention also a result of Dawson and Ivanoff [21] saying that in case $d \geq 3$ the random field converges to a nontrivial infinitely divisible field as $t \to \infty$. Similar results with different models are given by Cox [16], Kallenberg [29] and Durrett [25] among others.

The clustering, non–clustering phenomenon, briefly mentioned in Chapter 8 is studied in detail by Dawson [20].

Finally we mention a few questions not treated in the present monograph but having a rich literature.

(i) In case of critical branching random field one can investigate the properties of $\Lambda(A,t)$ when t is fixed but A is a big subset of $\mathbb{Z}^d$. We refer to Ivanoff [27].

(ii) Branching random walk in random environment. Here the fertility of a particle (i.e. the distribution $\{p_k\}$) depends on the temporary location of the particle. See [10] by Baillon et al.

(iii) Historical process. Knowing the location of a particle at time t what can be said on its past. See Dawson and Perkins [22].

Part III

A huge number of mathematicians devoted great effort to generalize the celebrated paper [53] of Strassen. Note that any point of the domain $D = \{(x,y) : y^2 \leq x,\ 0 \leq x \leq 1\}$ is a point of some functions of the Strassen class S. LePage and Schreiber [33, 34, 35] proved that, having an increasing number of independent Wiener processes, any point of D will be approximated by one of the normalized

$w_T(x)$ (cf. Section 10.1) of the underlying Wiener processes. [23] by Deheuvels and Révész was influenced by the mentioned three papers of LePage and Schreiber.

The results of Section 10.2 are due to Deheuvels and Révész [23]. In fact this paper considers only the case $d = 1$, otherwise the results are much more general.

Theorems 11.2, 11.4 and 11.5 give the shape of the domain where the particles are located at time T. This question was investigated earlier in several papers. Here we only mention [11, 12, 13] by Biggins.

References

1. R. J. Adler, R. E. Feldman and M. Lewin: Intersection local times for infinite systems of Brownian motions and for the Brownian density process, *The Annals of Probability* **19** (1991) 192–220.

2. R. J. Adler and J. S. Rosen: Intersection local times of all orders for Brownian and stable density processes—construction, renormalisation and limit laws, *The Annals of Probability* **21** (1993) 1073–1123.

3. S. Asmussen and H. Hering: *Branching Processes* (Birkhäuser, Boston, 1983).

4. S. Asmussen and N. Kaplan: Branching random walks I, *Stochastic Processes and their Applications* **4** (1976) 1–13.

5. S. Asmussen and N. Kaplan: Branching random walks II, *Stochastic Processes and their Applications* **4** (1976) 15–31.

6. K. B. Athreya and P. E. Ney: *Branching Processes* (Springer–Verlag, Berlin, 1972).

7. K. B. Athreya and P. E. Ney: Limit theorems for the means of branching random walks, in *Transactions of the sixth Prague conference*, 63–72, ed. J. Kožešnik (Academia Publ. House of the Czechoslovak Acad. of Sci. Prague, 1973).

8. P. Auer and K. Hornik: On the number of points of a homogeneous Poisson process, *J. of Multivariate Analysis* **48** (1994) 115–156.

9. P. Auer, K. Hornik and P. Révész: Some limit theorems for the homogeneous Poisson process, *Statistics & Probability Letters* **12** (1991) 91–96.

10. J. B. Baillon, Ph. Clément, A. Greven and F. den Hollander: A variation approach to branching random walk in random environment, *The Annals of Probability* **21** (1993) 290–317.

11. J. D. Biggins: The first– and last–birth problems for a multitype age–dependent branching process, *Advances of Applied Probability* **8** (1976) 446–459.

12. J. D. Biggins: Chernoff's theorem in the branching random walk, *Advances of Applied Probability* **14** (1977) 630–636.

13. J. D. Biggins: The asymptotic shape of the branching random walk, *Advances of Applied Probability* **10** (1978) 62–84.

14. J. D. Biggins and N. H. Bingham: Large deviations in the supercritical branching process, *Advances of Applied Probability* **25** (1993) 757–772.

15. M. Bramson, P. Ney and J. Tao: The population composition of a multitype branching random walk, *The Annals of Applied Probability* **2** (1992) 575–596.

16. J. T. Cox: Some remarks on the theory of critical branching random walk, in *Progress in Probability, Spatial stochastic processes*, 23–33, ed. K. S. Alexander and J. C. Watkins (Birkhäuser, Boston, 1991).

17. J. T. Cox and D. Griffeath: Large deviations for Poisson systems of independent random walks, *Z. Wahrscheinlichkeitstheorie verw. Gebiete* **66** (1984) 543–558.

18. J. T. Cox and D. Griffeath: Occupation times for critical branching Brownian motions, *The Annals of Probability* **13** (1985) 1108–1132.

19. M. Csörgő and P. Révész: *Strong Approximation in Probability and Statistics* (Academic Press, New York, 1981).

20. D. A. Dawson: Measure–valued Markov processes, in *Ecole d'Eté de Probabilités de Saint-Flour, 1991. Lecture Notes in Math.*, 1–260, ed. P. L. Hennequin (Springer–Verlag, Berlin, 1993).

21. D. Dawson and G. Ivanoff: Branching diffusions and random measures, in *Branching Processes*, 61–103, ed. A. Joffe and P. Ney (Marcell Dekker, New York, 1978).

22. D. A. Dawson and E. A. Perkins: Historical processes in *Memoirs Amer. Math. Soc.* no. 454 (1991).

23. P. Deheuvels, and P. Révész: On the coverage of Strassen-type sets by sequences of Wiener processes, *Journal of Theoretical Probability* **6** (1993) 427–449.

24. J. L. Doob: *Stochastic processes* (J. Wiley, New York, 1953).

25. R. Durrett: An infinite particle system with additive interactions, *Advances of Applied Probability* **11** (1979) 355–383.

26. J. Fleischman: Limiting distributions for branching random fields, *Transactions of the American Mathematical Society* **239** (1978) 353–389.

27. G. Ivanoff: The branching random field, *Advances of Applied Probability* **12** (1980) 825–847.

28. T. E. Harris: *The Theory of Branching Processes* (Springer–Verlag, Berlin, 1963).

29. O. Kallenberg: Stability of critical cluster fields, *Mathematische Nachrichten* **77** (1977) 7–43.

30. H. Kesten and B. P. Stigum: A limit theorem for multidimensional Galton–Watson processes, *Annals of Mathematical Statistics* **37** (1966) 1211–1223.

31. H. Kesten and B. P. Stigum: Additional limit theorems for indecomposable multidimensional Galton–Watson processes, *Annals of Mathematical Statistics* **37** (1966) 1463–1481.

32. G. F. Lawler: *Intersections of Random Walks* (Birkhäuser, Boston, 1991).

33. R. LePage and B. M. Schreiber: An iterated logarithm law for families of Brownian paths, *Z. Wahrscheinlichkeitstheorie verw. Gebiete* **70** (1985) 341–344.

34. R. LePage and B. M. Schreiber: A square root law for diffusing particles, in *Probability in Banach Spaces V* (Proc. Conf. Medford, Massachusetts). *Lecture Notes in Mathematics* **1153** (Springer–Verlag, Berlin, 1984, 1985).

35. R. LePage and B. M. Schreiber: Addendum to: An iterated logarithm law for families of Brownian paths, *Probability Theory and Related Fields* **76** (1987) 629.

36. T. M. Liggett: *Interacting Particle Systems* (Springer–Verlag, Berlin, 1985).

37. C. J. Mode: *Multitype Branching Processes* (Elsevier, New York, 1970).

38. J. E. Moyal: Discontinuous Markov process, *Acta Math.* **98** (1967) 221–264.

39. P. Ney: Branching random walk, in *Progress in Probability, Spatial stochastic processes* 3–22, ed. K. S. Alexander and J. C. Watkins (Birkhäuser, Boston, 1991).

40. S. C. Port: A system of denumerably many transient Markov chains, *The Annals of Mathematical Statistics* **36** (1966) 406–411.

41. S. C. Port: Equilibrium systems of recurrent Markov processes, *J. Math. Anal. Appl.* **18** (1967) 345–354.

42. S. C. Port, C. J. Stone and N. A. Weiss: SLLNs and CLTs for infinite particle systems, *The Annals of Probability* **3** (1975) 753–761.

43. P. Révész: How random is random? *Probability and Mathematical Statistics* **4** (1984) 109–116.

44. P. Révész: *Random Walk in Random and Non-Random Environments* (World Scientific, Singapore, 1990).

45. P. Révész: Black holes on the plane drawn by a Wiener process, *Probability Theory and Related Fields* **93** (1992) 21–37.

46. P. Révész: Path properties of an infinite system of Wiener processes, *Journal of Theoretical Probability* **6** (1993) 353–383.

47. P. Révész: Strong theorems on the extreme values of stationary Poisson processes, To appear.

48. P. Révész: The distribution of the particles of a branching random walk, To appear.

49. S. Sawyer: Branching diffusion processes in population genetics, *Advances of Applied Probability* **8** (1976) 659–689.

50. A. V. Skorohod: Branching diffusion process, *Theor. Probability Appl.* **9** (1964) 492–497.

51. F. Spitzer: Electrostatic capacity, heat flow, and Brownian motion, *Z. Wahrscheinlichkeitstheorie verw. Gebiete* **3** (1964) 110–121.

52. U. Stadtmüller: A note on the law of iterated logarithm for weighted sums of random variables, *The Annals of Probability* **12** (1984) 35–44.

53. V. Strassen: An invariance principle for the law of the iterated logarithm, *Z. Wahrscheinlichkeitstheorie verw. Gebiete* **3** (1964) 211–226.

54. K. Uchiyama: Spatial growth of a branching process of particles living in $\mathbb{R}^d$, *The Annals of Probability* **10** (1982) 896–918.

55. J. B. Walsh: An introduction to stochastic partial differential equations, in *École d'Été de Probabilités de Saint-Flour XIV-1984. Lecture Notes in Math.* **1180**, 265–439 (Springer–Verlag, Berlin, 1986).

56. N. A. Weiss: Limit theorems for infinite particle systems, *Z. Wahrscheinlichkeitstheorie verw. Gebiete* **20** (1971) 87–101.

57. N. A. Weiss: The occupation time of a set by countably many recurrent random walks, *Ann. Math. Statist.* **43** (1972) 293–302.

Author Index

Subject Index

www.ingramcontent.com/pod-product-compliance
Lightning Source LLC
Chambersburg PA
CBHW050640190326
41458CB00008B/2358
* 9 7 8 9 8 1 0 2 1 7 8 4 6 *